Bacteria B...   :ctric

Isabella Ava

Copyright © 2023 by Isabella Ava

All rights reserved. No part of this book may be reproduced in any manner whatsoever without written permission except in the case of brief quotations embodied in critical articles and reviews.

First Printing, 2023

Table of Contents

Chapter	Page No.
Chapter 1: Preamble and Scope of Research	1-13

1.1 Introduction
    1.1.1 Microbial Electrochemical Systems
1.2 Electron transport in microorganisms
1.3 Specific EET in Metal Respiring Bacteria
1.4 Bioanode in MES- Electrode reducing bacteria
1.5 Biocathode in MES- Electrode oxidizing bacteria
1.6 Research problem statement
1.7 Scope of Research
1.8 Objectives of the research

Chapter 2: Present State of Knowledge	14-32

2.1. Introduction
    2.1.1 Microbial Fuel cells (MFC)
    2.1.2 Bioelectrochemical synthesis (BES)
    2.1.3 Microbial electrolysis cell (MEC)
2.2 Biocatalyst in MES
2.3 Types of biocatalyst used in METs
    2.3.1 Bacteria
        2.3.1.1 Electrode-Microbe interactions
        2.3.1.2 Electron transport across cell membrane
        2.3.1.3 Extracellular electron transport- gram negative bacteria
        2.3.1.4 MTR pathway in *Shewanella*
        2.3.1.5 Nanowires- *Geobacter*
        2.3.1.6 Extracellular electron transfer- gram positive bacteria
    2.3.2 Yeast
    2.3.3 Fungi
    2.3.4 Algae
        2.3.4.1 Algal bioanode
        2.3.4.2 Algal biocathode
    2.3.5 Photosynthetic bacteria
    2.3.6 Cyanobacteria
2.5 Quorum sensing
2.6 Mixed culture biocatalyst for MES application

Table of Contents

Chapter 3: Enhancing extracellular electron flux by overexpression of NADH dehydrogenase II towards improving bioelectrogenic activity of MES   33-61

3.1 Introduction

3.2 Experimental

    3.2.1 Strains and growth conditions

    3.2.2 Cloning and expression of NDH-2

       3.2.2.1 General DNA techniques

       3.2.2.2 Isolation of plasmid DNA

       3.2.2.3 Restriction digestion of DNA samples

       3.2.2.4 Agarose gel electrophoresis

       3.2.2.5 Gel purification of DNA fragments

       3.2.2.6 Polymerase chain reaction

       3.2.2.7 Ligation and double digestion

       3.2.2.8 DNA sequencing

       3.2.2.9 Preparation of competent cells and transformation

    3.2.3 Colony PCR

    3.2.4 Protein purification

    3.2.5 Bioelectrochemical system design and characterization

    3.2.6 Bioelectrochemical analysis

    3.2.7 Estimation of biofilm on the anode by biofilm viability kit

    3.2.8 NAD/NADH estimation

    3.2.9 Iron utilization assays using Ferrozine assay

    3.2.10 Biohydrogen production

    3.2.11 Spectro-electrochemical studies

3.3 Results and Discussion

    3.3.1 Cloning and Protein purification

    3.3.2 Bioelectrogenesis of NDH-2 clone

    3.3.3 Chronoamperometry (CA)

    3.3.4 Cyclic voltammetry (CV)

    3.3.5 Potentiostatic electrochemical impedance spectroscopy (PEIS)

    3.3.6 Effect of NDH-2 on NAD/NADH concentration, bioelectricity and biofilm development

    3.3.7 Biohydrogen production

    3.3.8 Iron reduction by NDH-2 expressing cells

    3.3.9 Spectro-electrochemical studies on purified NDH-2

3.4 Conclusions

Chapter 4: Development of Electrogenic Consortia by Selective Enrichment Strategies and Application for Real-field Scenario   62-119

4.1 Introduction

4.2 Selective enrichment strategies for electrogenic consortia: Effect of Heat and Iodopropane pretreatment

4.2.1 Experimental

    4.2.1.1 Biocatalyst and Selective enrichment methods

    4.2.1.2 Experimental design and operation

    4.2.1.3 Analysis

4.2.3 Results and discussion

    4.2.3.1 Bioelectrogenesis- Pretreated Vs Untreated

    4.2.3.2 Polarization behavior

    4.2.3.3 Coulombic efficiency (CE) and substrate degradation

    4.2.3.4 Cyclic voltammetry

    4.2.3.5 Chronoamperometry

    4.2.3.6 pH and VFA

4.3 Selective enrichment strategies for electrogenic consortia: Effect of acid and BESA pretreatment

4.3.1 Experimental- Biocatalyst pretreatment

4.3.2 Results and discussion

    4.3.2.1 Bioelectrogenesis

    4.3.2.2 Polarization

    4.3.2.3 Columbic efficiency and substrate degradation

    4.3.2.4 Cyclic Voltammetry

    4.3.2.5. Chronoamperometry (CA)

    4.3.2.6 pH and VFA variations

4.3.3 Conclusion

4.4 Metagenomic analysis to understand functional relation between microbial dynamics with enrichment.

4.4.1 DNA extraction

4.4.2 MiSeq analysis

4.4.3 Microbial Dynamics- heat and Iodopropane enrichment

    4.4.3.1 Sequence analysis and diversity estimation

    4.4.3.2 Diversity of communities

    4.4.3.3 Conclusions

4.4.4 Microbial dynamics - BESA and acid pretreatment in comparison to heat and Iodopropane enrichment

    4.4.4.1 Species richness indicators

    4.4.4.2 Diversity of communities

    4.4.4.3 Conclusions

4.5 Microbial electrochemical systems for bioelectricity production with simultaneous treatment of paper and pulp wastewater

4.5.1 Experimental

    4.5.1.1 Paper Mill Effluent

    4.5.1.2 Biocatalyst

    4.5.1.3 MFC system design and operation

    4.5.1.4 Analysis

4.5.2 Results and Discussion

    4.5.2.1 Bioelectrogenic activity- MFC Vs AnT

        4.5.2.1.1 Chronoamperometry

        4.5.2.1.2 Cyclic voltammetry

        4.5.2.1.3 Electron Transfer

        4.5.2.1.4 First derivative

    4.5.2.2 Fuel cell behaviour

        4.5.2.2.1 Bioelectrcity

        4.5.2.2.2 Polarization

        4.5.2.2.3 Cell potentials

        4.5.2.2.4 Relative decrease in anode potential (RDAP)

    4.5.2.3 Bioremediation-MES Vs AnT

    4.5.2.4 Integrated process for color removal

    4.5.2.5 VFA and pH variation

4.5.3 Conclusions

Chapter 5: Summary and Future Perspectives

5.1 Summary and Conclusion of Research

5.2 Future perspective

Abbreviations

Weights and measures

bp, kb	Base pair, kilobase
°C	Degree centigrade
ng, µg, mg, g	Nanogram, microgram, milligram, gram
sec, min, h	Second, minute, hour
kDa	Kilodalton
V, KV	Volts, Kilovolts
mA, mV	Milliampere, millivolt
µl, ml, L	Microliter, milliliter, liter
pmol, nmol, mmol	Picomole, nanomole, millimole
nM, µM, mM, M	Nanomolar, micromolar, millimolar, molar
A$_{xxx}$	Absorbance or optical density where xxx stands for the wavelength of light at which the absorbance is measured
OD	Optical density
%	Percent
rpm	Revolutions per minute

Symbols

α	Alpha
β	Beta
γ	Gamma
δ	Delta
~	Approximately

Techniques

SDS–PAGE	Sodium dodecyl sulphate–polyacrylamide gel electrophoresis
PCR	Polymerase chain reaction
CV	Cyclic Voltammetry
PIES	Potentiostatic Electrochemical Impedance Spectroscopy
CA	Chronoamperometry

Chemicals

APS	Ammonium persulphate
ATP	Adenosine-5'-triphosphate
BSA	Bovine serum albumin
dNTP	2'-deoxynucleotide-5'-triphosphate
PBS	Phospate buffered saline

Abbreviations

IPTG	Isopropyl- β-D- thiogalactopyranoside
PMSF	Phenylmethylsulfonyl fluoride
TEMED	N,N,N',N'-tetramethyl-ethane-1,2-diamine
Tris	Tris(hydroxymethyl)aminomethane
DDM	n-Dodecyl β-D-maltoside
DMSO	Dimethyl sulfoxide

Miscellaneous

PVDF	Polyvinylidene fluoride
IDP	Iodopropane
BESA	2-Bromoethanesulfonic Acid
CE	Coulombic efficiency
MTR pathway	Metal reducing pathway
ETC	Electron transport chain
EET	Extracellular electron transport
EAB	Electroactive bacteria
NAD	Nicotinamide adenine dinucleotide
NADH	Nicotinamide adenine dinucleotide, Reduced
FAD	Flavin adenine dinucleotide
$FADH_2$	Reduced flavin adenine dinucleotide
FMN	Flavin Mononucleotide
OMCs	Outer membrane cytochromes
OmcZ	Octaheme cytochrome
PpcA	Trihemeperiplasmic c-type cytochrome
MQ	Menaquinones
H_2ase	Hydrogenase
QH_2	Hydroquinol
Q	Quinone.
MQH2	Reduced menaquinones
MacA	Metal-reduction-associated cytochrome
EMOx	Electron mediator oxidized
EMRed	Electron mediator Reduced
hυ	Light source
PS II	Photosystem II
PS I	Photosystem I
ATP	Adenosine triphosphate
ADP	Adenosine diphosphate
VFA	Volatile fatty acids
CDP	Cell design point
RDAP	Relative decrease in anode potential

Abbreviations

AnT	Anaerobic treatment
COD	Chemical oxygen demand
TEA	Terminal electron acceptor
SHE	Standard hydrogen electrode
DET	Direct electron transport
MET	Mediated electron transport

List of Figures

Fig 1.1	Extracellular electron transport and its application in various microbial electrochemical systems	3
Fig 1.2	Role of biocatalyst at bioanode: Various protein-cytochrome complexes involved in extracellular electron transport across other membrane in bacteria (*Geobacter sulfurreducens*).	7
Fig 1.3	Role of biocatalyst at biocathode: Proposed mechanism of electron entry form external environment or cathode into the *Shewanella onediensis*.	9
Fig 2.1	Schematic diagram of MES and its applications.	15
Fig 2.2	Schematic diagrams of single and dual chambered microbial fuel cells	16
Fig 2.3	Applications and scope of microbial electrochemical system and the wide range of substrates they can use as substrates.	18
Fig 2.4	Various biocatalysts used in microbial electrochemical systems.	19
Fig 2.5	Standard redox potentials of various redox proteins and molecules of electron transport chain interacting with electrode (Standard redox potentials are calculated at pH 7 and 25°C Vs SHE)(Milo and Phillips., 2015).	21
Fig 2.6	Modes of extracellular electron transport in microbial electrochemical systems a) direct b) nanowires and c) mediators.	22
Fig 2.7	Role of biocatalyst at bioanode: Extracellular electron transport in yeast cells. In yeast cell, electrons are transported to electrode by soluble redox mediators which can be metabolically generated or from externally source.	27
Fig 2.8	Role of algal biocatalyst at bioanode and respective extracellular electron transport.	28
Fig 2.9	Role of biocatalyst in biocathode: Microalgae at biocathode. Oxygen produced by the microalgae will accept the electrons as cathode enhancing the terminal step.	29
Fig 3.1	Mechanism of NADH dehydrogenase-II [It accepts electrons from NADH and donates them to quinone cycle without pumping protons from cytosol to periplasmic space].	36
Fig 3.2	Regulation of *ndh* gene (*E. coli*) and its expression (Keseler et al., 2017).	37

List of Figures

Fig 3.3	Electrodes and reactor setup used for the bioelectrochemical studies.	43
Fig 3.4	Three electrode electrochemical setup used for the study.	44
Fig 3.5	Double digestion of insert and vector for confirmation of insertion of gene into vector and Clones conformation using colony PCR for positive colonies	46
Fig 3.6	Cell pellets of NDH-2 induced (1 mM IPTG) and uninduced cultures. [Green color of the pellet is due to the presence of FAD cofactor in the NDH-2 enzyme].	48
Fig 3.7	Bound NDH-2 to the Ni-NTA affinity purification column after binding step [Color of the column is due to the FAD cofactor].	48
Fig 3.8	Protein purification using his tag purification (Ni-NTA beads). Purified protein was light greenish in color emphasizing the presence of FAD^+ cofactor in the NDH-2.	49
Fig 3.9	Western blotting of the NDH-2 protein using anti-6Xhis antibodies. Presence of band at ~42 KDa indicates the expression of the NDH-2 in the membrane fraction.	49
Fig 3.10	Absorption (visible) spectra of purified NDH-2 under oxidizing (air) and under reducing conditions (reduced by sodium dithionite and NADH). Addition of dithionite has resulted in reduction of FAD^+ which lead to the over decrease in absorbance at 450 nm. Charge transfer complex (CTC) formation can only be seen in case of NADH.	50
Fig 3.11	Bioelectrogenic activity of NDH-2 induced and uninduced BES reactors poised at + 0.2 V. NDH-2 induced culture has shown 3 fold increases in current production compared to uninduced MES reactor.	51
Fig 3.12	Cyclic voltammograms recorded during 48 hours period of time (at every six hours) at a scan rate of 0.001 V. s^{-1}(a) NDH-2 induced (b) Uninduced (c) Combined.	53
Fig 3.13	Biofilm monitoring by EIS (a) NDH-2 induced (b) Uninduced (c) Combined. Higher charge transfer decrement was observed in case of NDH-2 induced cells.	55
Fig 3.14	Biofilm formation on the FTO plates after impedance monitoring. Biofilm was stained using SYTO9 and propidium iodide dyes. After 48 hour cycle, it was observed that the cell density was more on the NDH-2 induced MES.	56

List of Figures

Fig 3.15	Total NAD^+, NAD^+, NADH concentration of aerobically grown cultures after 3 hours of NDH-2 induction. $NADH/H^+$ concentrations were found to be very low corresponding to the high rate of catabolism and increased electron flux into electron transport chain.	57
Fig 3.16	Percentage hydrogen of total biogas produced in NDH-2 induced and uninduced cells after 48 hours of growth.	58
Fig 3.17	Iron reduction by NDH-2 and uninduced cells using Ferrozine assay with respect to time.	59
Fig 3.18	UV visible spectra of reduced NDH-2 when slowly oxidized using Potentiostat at a scan rate of 1 $mV.s^{-1}$.	60
Fig 4.1	Schematic diagram of pretreatment strategy used for enrichment of electroactive bacteria.	65
Fig 4.2	Variation in current density and power density of MFCs with respect to time.	69
Fig 4.3	Coulombic efficiency of iodopropane and heat pretreated MFCs with respect to untreated MFC. Untreated MFC has shown highest COD removal but low CE indicating low substrate to power conversion	71
Fig 4.4	Cyclic voltammograms of iodopropane, heat and untreated MFC during various hours of reactor operation.	72
Fig 4.5	Chronoamperometry curves depicting the sustainable current. CA was performed at an applied potential of 0.5 V for 30 minutes. Iodopropane showed highest sustainable current generation.	74
Fig 4.6	Feed with pH 6 was feed to the reactor initially and change of pH was monitored.	75
Fig 4.7	Production of metabolic acids or volatile fatty acids during MFC operation in pretreated and untreated MFCs.	75
Fig 4.8	Open circuit profile and current Profile with 100 Ω resistor of acid and BESA pretreated MFC with respect to the control.	77
Fig 4.9	Polarization curves of all the pre-treated fuel cells with respect to untreated fuel cell.	79
Fig 4.10	COD removal efficiency and coulombic efficiency of acid and BESA pretreated MFC with respect to untreated MFC.	80
Fig 4.11	Cyclic Voltammogram profiles of acid, BESA and control MFCs at various scan rates scan rates from in the scan range of 0.5 V to – 0.5 V.	82

List of Figures

Fig 4.12	Chronoamperometric profiles of pretreated MFC compared to untreated control MFC.	84
Fig 4.13	Rarefaction curves constructed using Unifrac.	88
Fig 4.14	Principal component analysis (PCA) for the 4 samples studied.	89
Fig 4.15	Relative abundance of various phyla in heat, Iodopropane, crude and control samples.	90
Fig 4.16	Abundance of various classes of proteobacteria in all the four samples and their distribution.	91
Fig 4.17	Pathway for methane synthesis in methanogens and its inhibition by Iodopropane (Kenealyet al., 1981).	92
Fig 4.18	Principal component analysis (PCA) and rarefraction curves for the 4 pretreated MES reactors compared to control and parent inoculums.	94
Fig 4.19	Relative abundance of various phyla in acid, BESA, heat, Iodopropane, crude and control samples.	95
Fig 4.20	Relative abundances of various classes of in all the six samples and their distribution.	96
Fig 4.21	Chrono-amperometry depicting the production of sustainable current during the MES and AnT operation	102
Fig 4.22	Cyclic voltammograms showing variation in the electron discharge pattern (+0.5 to -0.5 V) with anode and cathode as working and counter electrodes against Ag/AgCl (S) reference electrode (scan rate, 30 mV/s).	103
Fig 4.23	Tafel slopes derived with the function of biosystems, pH and time.	105
Fig 4.24	Derivative cyclic voltammograms (DCV) recorded during operation of MES and AnT systems with respect to different pH and time.	107
Fig 4.25	Progressive increase in OCV and current density generation during MFC operation with paper and pulp wastewater.	108
Fig 4.26	Polarization curves (Z1 is zone of activation losses; Z2 is zone of ohmic losses; Z3 is zone of concentration losses) of MFC operated at pH 6 and pH 7.	109
Fig 4.27	Cell potentials and relative decrease in anodic potential (RDAP) profile drawn against varying external load during stabilized performance of MFC with the function operating pH.	110

List of Figures

Fig 4.28	Relative decrease in anodic potential (RDAP) profile drawn against varying external load during stabilized performance of MFC with the function operating pHs.	112
Fig 4.29	Variation in phosphates, sulphates and colour removal pattern observed with the function of time during the operation of MES and AnT systems (HRT, 48 hrs; organic loading rate, 3000 mg/l).	113
Fig 4.30	Variation in COD observed with the function of time during the operation of MES and AnT systems (HRT, 48 hrs; organic loading rate, 3000 mg/l).	114
Fig 4.31	Color removal efficiency of MFC and AnT reactors.	114
Fig 4.32	Variations of pH documented during MFC and AnT system operation at with the function of pH and time	116
Fig 4.33	Variations of VFA documented during MFC and AnT system operation at with the function of pH and time.	117

List of Tables

Table 3.1	Primers used for cloning and expression of ndh gene in this study	38
Table 3.2	Buffers and solutions used for agarose gel electrophoresis of DNA samples.	39
Table 3.3	Buffers and solutions used for SDS-PAGE analysis of proteins.	41
Table 3.4	Buffers and solutions used for western blotting.	42
Table 3.5	Properties of NDH-2 protein based on protein sequence information (using ExPASy)	47
Table 4.1	Redox currents obtained from cyclic voltammograms during various hours of operation at 30mV/s scan rate.	82
Table 4.2	Peak potentials identified in cyclic voltammograms.	83
Table 4.3	Comparison of various diversity parameters of 16s rRNA sequences obtained by MiSeq.	88

CHAPTER 1
Preamble and Scope of Research

1.1 Introduction

1.1.1 Microbial Electrochemical Systems (MES)

Microbial electrochemical systems (MES) use biocatalyst to convert the chemical energy present in the organic electron donors to bioelectricity and other biobased chemicals. The principle of MES is the oxidation of organic or inorganic electron donors by the self-sustaining electroactive microorganisms known as biocatalyst at the anode. Insertion of electrodes in MES will allow the electroactive microorganism to donate electrons from its oxidative metabolism to electrodes. Connection of anode to cathode via external circuit will result in directional movement of electron from anode to cathode due to potential difference. At cathode, they will be accepted by either by abiotic electron acceptor like oxygen, nitrates etc or biotic component like bacteria or algae depending on the application of the MES. Microbial electrochemical systems can be classified in to following systems based on the function and application of the system (Fig 1.1). Microbial fuel cells (MFC) are used for simultaneous biodegradation of organic wastes and harvest bioelectricity, microbial electrolysis cells (MEC) for hydrogen production, microbial desalination cells (MDC) for synchronous wastewater treatment and seawater desalination, and microbial electrosynthesis (MES) for production of valuable chemicals and biofuels from CO_2 bio-reduction and electrofermentation (EF) for conversion of organic wastes and biomass for production of platform chemicals.

MESs have emerged as a potential alternative to conventional energy sources for sustainable production of energy and value added products using microorganisms as biocatalyst in biologically induced electrogenic microenvironment. The concept of remodeling microbial metabolism to recover energy and value added products by degradation of organic matter makes MES a flexible and sustainable platform (Venkata Mohan et al., 2014). Microbes generally carry their metabolic activities by utilizing available substrate and generates the reducing equivalents [protons (H^+) and electrons (e^-)] which move via a series of redox components/carriers (NAD^+, FAD, FMN, etc.) towards an available terminal electron acceptor (TEA) and generates a proton motive force that facilitates generation of high energy phosphate bonds (adenosine tri phosphate (ATP) which are useful for the microbial growth and subsequent metabolic activities). Electrons (e^-) in the reduced substrate (eg. carbohydrates, organic acids, H_2S, CH_4) initiate an "electric circuit" that ends when the electrons reach the e^- sink furnished by a final electron acceptor (Venkata Mohan et al., 2013b). The function of microbial electrochemical system (MES) is based on harnessing the available electrons by artificially introduced electrodes as intermediary electron acceptors. Anode chamber, a biofactory facilitates the growth of bacteria and helps in generation of reducing equivalents required for bio-electrochemical redox reactions under anaerobic microenvironment. MES facilitates direct conversion of chemical energy to value added products or energy (Cheng and Logan 2007; Rabaey and Rozendal, 2010; Venkata Mohan et al., 2007, 2008, 2013a, b). At present they are being used to harvest bioelectricity, to remediate complex pollutant in wastewater, to produce various forms of value added products, etc., either in the anode chamber, cathode chamber or in the combination of both (Venkata Mohan et al., 2013b). MES integrates biodegrdation, electrolytic dissociation and electrochemical oxidation/reduction processes in the presence of the solid electron acceptor (anode and cathode) during metabolism (Wang and Ren, 2013; Li et al., 2013; Huang et al., 2011; Venkata Mohan et al., 2009a). Essential function of all MES depends on the electron transfer mechanism and cascade of redox reactions in the absence of oxygen.

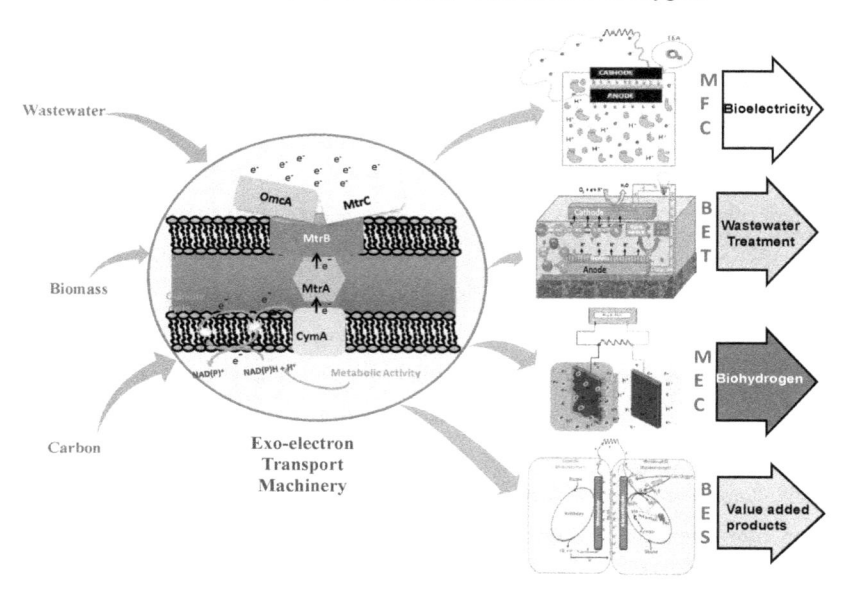

Fig 1.1: Extracellular electron transport and its application in various microbial electrochemical systems

Among MESs, MFC garnered significant interest in both basic and applied research due to its potential to harness clean energy (Venkata Mohan et al., 2007, 2008). In MFC, the oxidation reaction (generating the reducing equivalents) is separated from terminal reduction reaction by a selectively permeable ionic membrane (proton exchange membrane) which facilitate in capturing of electrons anode. MFC has shown potent to extract electric current from a wide range of soluble or dissolved complex organic wastes/wastewater and renewable biomass and thus may offset the operational cost of effluent treatment (Pant et al., 2012; Venkata Mohan et al., 2013a,b). They are viewed as viable and effective treatment units in addition to bioelectricity generation and therefore can be termed as bio-electrochemical treatment (BET) system when the primary focus is towards treatment (Venkata Mohan et al., 2009a, 2011; Velvizhi and Venkata Mohan, 2011). Product recovery is another alternative, where external applied potential is used to produce various value added products through microbial electrolysis cell (MEC) and microbial electrosynthesis process (Cheng and Logan 2007; Rabaey and Rozendal, 2010; Lenin Babu et al., 2013; Venkata Mohan and Lenin Babu, 2011). In the case of MEC, the H^+ produced in the anode chamber migrates to the cathode and gets reduced to form H_2 in presence of electrons arrived from the anode. A small voltage is applied to the system which is required to cross the endothermic barrier to form H_2 gas. Other than H_2, various value added products can be recovered through bio-electrosynthesis or microbial electrosynthesis (Rabaey and Rozendal, 2010). With the aid of small input of electric power, many value added compounds like ethanol, butanol, glutamic acid, succinate etc., can be synthesized at the cathode (Steinbusch et al., 2010).

1.2 Electron transport in microorganisms

In bacterial electron transport chains (ETC), electrons are transported from a low potential electron donor to an acceptor with more positive redox potential by redox reactions. These electron transfer reactions are usually catalysed by membrane-bound proteins and redox carriers compounds that use the energy (potential energy) difference between electron donor and acceptor to establish a proton (sodium in few cases) gradient across the plasma or cellular membrane which is finally used for ATP synthesis. ETC thus converts the difference in electrical potential between donor and acceptor into chemical energy for the cell (Anraku, 1988).

In order to adapt to varying environmental conditions and gain maximum energy available microorganisms have developed an enormous diversity of electron transport chains (ETC) (Hernandez and Newman, 2001). ETCs primarily consist of primary dehydrogenases that accept high energy electrons from donors such as NADH, $FADH_2$, etc. They usually couple electron transport to H^+ or Na^+ transport across the membrane (Anraku, 1988). Also involved in transmembrane ion-transport are membrane-localized (multi-) protein complexes such as cytochromes and terminal oxidases (reductases) that transfer electrons to a final acceptor such as oxygen, nitrate or fumarate (Hannemann et al., 2007; Vignais and Billoud, 2007; Richter et al., 2012). Most transmembrane reductases and oxidases function as ion pumps but some do not like type II NADH dehydrohgenases. Electron carrying co-factors such as quinones, flavins, heme, iron–sulfur clusters or copper ions also play an important role in microbial electron transport. Some of these are soluble lipophilic molecules that shuttle electrons between the relatively large enzymatic complexes inside the membrane (e.g., quinones) while others are catalytic cofactors bound to proteins (e.g., heme groups of cytochromes; Deller et al., 2008; Dolin, 2012). There are also membrane bound complexes that use the exergonic electron bifurcation of soluble cellular redox compounds such as ferredoxin (Fd) and NADH for transmembrane ion transport and therefore establishment of a motive force across the membrane that can drive ATP synthesis (Herrmann et al., 2008; Müller et al., 2008; Schuchmann and Müller, 2012).

The achievable energy gain (Gibbs free energy, ΔG) of each electron transport chain depends on the redox potential difference (ΔE) of all reactions between electron donor and acceptor. Some bacteria incorporate several electron transport chains, which they can use sometimes even simultaneously in order to respond to different electron acceptors and donors available in the environment (Haddock and Schairer, 1973; Anraku, 1988). Others are restricted to only one respiratory pathway (Das and Ljungdahl, 2003; Müller et al., 2008). This diversity of microbial electron transport mechanisms illustrates the complexity of the approach to bioelectrochemical techniques. In order to interfere efficiently with the redox metabolism of an organism one needs to understand the targeted site of EET and its metabolic impact.

A wide range of microbes has been discovered to be able to exchange electrons with solid surfaces (direct EET) and/or soluble mediators (indirect EET) but only a few have been studied in depth. In fact the mechanisms of electron transport that are found in different species can differ significantly from one another. Dissimilatory metal reducing bacteria are amongst the most studied being, able to respire insoluble metals in anaerobic environments. The model organisms *Geobacter sulfurreducens* and *Shewanella oneidensis* were studied by various research groups for decades and evidence for direct and indirect electron transfer between the organism and electrodes could be found (Bond and Lovley, 2003; Thrash and Coates, 2008; Rosenbaum et al., 2011; Ross et al., 2011). For both bacteria outer-membrane cytochromes were identified as essential compounds to enable EET (Mehta et al., 2005; Shi et al., 2007, 2009; Breuer et al., 2015). However, there are several differences in their electron transport chains, for example *Shewanella* excretes soluble electron carriers while similar compounds are missing in *Geobacter* sp. (Holmes et al., 2006; Marsili et al., 2008a). Furthermore it could be shown that the redox chains catalyzing an inward current rely on different mechanisms than current producing reactions (outward movement of electrons) (Dumas et al., 2008; Strycharz et al., 2008). Another group of dissimilatory metal reducing bacteria which is an obligate anaerobe *Thermincola*, was also found to be capable of directly transferring electrons to anodes (Marshall and May, 2009; Wrighton et al., 2011). Their EET mechanisms also seem to rely on cytochromes that in this case are cell-wall associated of the Gram-positive bacteria (Carlson et al., 2012). Interestingly there are also organisms such as *Clostridium ljungdahlii, Bacillus subtilis* which do not have any cytochromes but were tested positive on EET (Köpke et al., 2010; Nevin et al., 2011). The exact mechanisms by which electrons are transferred between the electrode surface and the microbial metabolism still remain unclear (Patil et al., 2012; Sydow et al., 2014).

1.3 Specific EET in Metal Respiring Bacteria

Metal reducing or respiring bacteria have the ability to interact with solid minerals, e.g., Fe(III) or Mn(IV), to obtain energy by using those minerals as electron acceptors and/or donors for their respiration process. Here a transport of electrons from a low redox potential donor to an acceptor with a higher redox potential can result in a proton gradient to drive ATP synthesis. This metal reducing feature of these bacteria plays an important role in biogeochemical cycles and has the potential to be used in bioremediation and MES (Lovley et al., 2004; Richardson et al., 2012).

In contrast to other respiration processes, where freely diffusible gas or readily soluble substances can easily enter the cell and be used as electron acceptors/donors, the major challenge for metal respiring bacteria is their interaction with the extracellular minerals, which cannot diffuse through the cell membrane and do not have access to the neither periplasm nor cytoplasm. To overcome this barrier these bacteria need redox proteins on their outer membrane or have to excrete redox-active shuttle molecules to transfer electrons between the cytoplasm and the extracellular metals (Hernandez and Newman, 2001; Hartshorne et al., 2009). There exists a great diversity of mechanisms for such electron-shuttling pathways, which is explained in more detail in introduction by describing the EET of organisms belonging to several groups of microorganisms.

1.4 Bioanode in MES- Electrode reducing bacteria

Electroactive bacteria growing on an anode have the ability to oxidize a wide range of organic substrates and reduce the electrode by means of EET. This phenomenon of electron transfer to electrode was understood with the finding that non-fermentable substrate such as acetate could be oxidized to carbon dioxide with direct transfer of electron to an electrode serving as a sole electron acceptor (Bond et al., 2002). This demonstrated that electron transfer to electrode is a respiratory process as there is no chance for substrate level phosphorylation with acetate as an electron donor under anaerobic conditions.

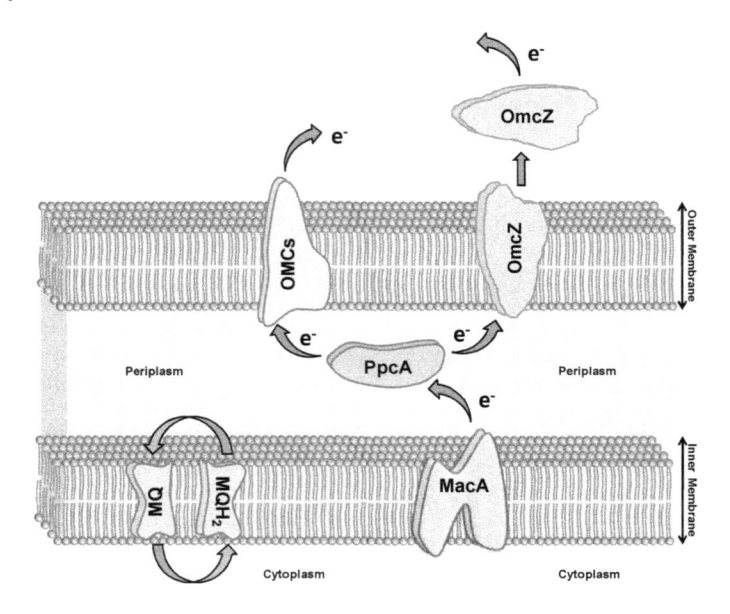

Fig 1.2: Role of biocatalyst at bioanode: Various protein-cytochrome complexes involved in extracellular electron transport across other membrane in bacteria (*Geobacter sulfurreducens*).

The outcomes of this and other related studies have proven that bacteria can extract electrons from the organic source and convert them to power with reasonably good efficiency. About 100 years ago evidence was shown that electrons obtained during the microbial metabolism can be harvested with electrodes, but then discovery of electroactive bacteria that oxidizes the organic matter and transfer the electrons to the anode is a real milestone (Debabov, 2008). The approach to identify the bacteria that contribute to power production from various sources is through identification of those bacteria that selectively colonize anode surfaces and those bacteria which can reduce the mineral ores. The colonization of specific bacteria on the anode surface depends on the enrichment conditions (Jung and Regan, 2007; Kim et al., 2007; Lee et al., 2008). Maximum potential difference that can be achieved by bacteria is +1.2 V, as the lowest redox potential of substrate that can be used in anode chamber is - 0.4 V (vs. SHE) and the oxidant with highest potential that can be used in cathode is oxygen (+0.8 V (vs. SHE)) giving a theoretical maximum voltage of +1.2 V. However, theoretical maximum voltage was calculated without taking internal losses into consideration leading to the lower values in the real conditions. Another limitation of biocatalyst at the anode is the rate of electron transfer to the anode which is limited by the metabolic rate of the biocatalyst and the distance of the biocatalyst from the anode. These limitations can be addressed by using genetic and metabolic engineering approaches such as cloning genes, which can increase the electron flux through the EET pathway. The power production capabilities of electroactive bacteria is understood by analyzing the rate of electron transfer to the anode at various voltages, anode potentials, electron donor concentrations, etc. and interpretation of this information helps to determine the biomolecular component involved in electron transfer to the anodes (Fricke et al., 2008).

1.5 Biocathode in MES- Electrode oxidizing bacteria

Unlike anode-respiring bacteria, microorganisms can also utilize the cathode as an electron donor in an MET, using electrode as an energy (electron) source (Fig. 1.3). Microorganisms that accept electrons directly from the cathode are referred to as electrode-oxidizing bacteria (Saratale et al., 2017). *Geobacter* was the first organism shown to use cathode as an electron donor. Only a limited number of known electrode-oxidizing bacteria, including *Clostridium ljungdahlii, C. aceticum, Shewanella oneidensis, G. metallireducens, G. sulfurreducens*, etc., have been reported (Choi and Sang, 2016) were studied till now. These bacteria are associated with the cathode for production of biofuels, including H_2, methane, formate, and various liquid organic fuels (Christodoulou et al., 2017). Organisms growing on a cathode can utilize carbon substrates such as acetate or CO_2 and use H_2 or formate generated in the cathode as an electron donor to produce methane and higher organic fuels such as 2,3-butanediol, butyrate, butanol, etc. It was also reported in several studies that the acetate can be converted to medium-chain fatty acids such as caproate and caprylate at the biocathode (Van Eerten-Jansen et al., 2013).

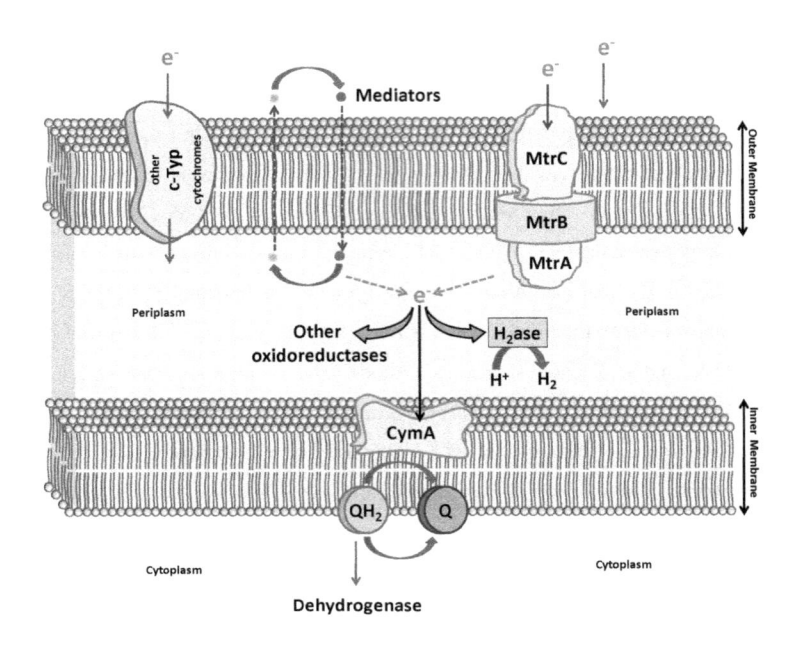

Fig 1.3: Role of biocatalyst at biocathode: Proposed mechanism of electron entry form external environment or cathode into the *Shewanella onediensis*. Mtr – Metal reducing; H$_2$ase – Hydrogenase; CymA - ; QH$_2$ – Hydroquinol; Q – Quinone.

Microorganisms are capable of coupling direct electron consumption to their metabolism for fuel production, but the efficient cathodic catalysis relies on the growth and association of multiple species with each one playing their role. Bacteria capable of using cathode as an electron donor are also known for mediating reduction of Uranium (VI), selenium, tellurium, and chlorinated compounds (Choi and Sang, 2016) which play important role in biogeochemical cycles and bioremediation by leaching the minerals and immobilizing the radionuclides (Lovley, 2013). Although several bacteria are involved in cathodic reactions, mechanism on the cathode side of a biocatalyst is not very well known.

Bacillus subtilis is one of the organisms which were shown to have EET but no defined EET mechanism has been discovered till now. Several studies with *Bacillus subtilis* have shown that the strain is capable of extracellular electron transport and was shown to generate significant voltages in both dual and single chamber configurations (Coman et al., 2009; Nimje et al., 2009). It is hypothesized that the native electron transport proteins might play role in EET in these bacteria. One such native electron transport protein which can involve in EET is type II NADH:quinone oxidoreductase or NADH dehydrogenase II or NDH-2. NDH-

2 was found to be a major NADH dehydrogenase in *Bacillus subtilis* which regulates the NADH/NAD$^+$ ratio in the cytoplasm and thereby the controlling overall metabolic state of the cell (Gyan et al., 2006). It plays vital role in both intracellular and extracellular electron transport by regulating the electron transfer flux into these pathways. It is a non-proton pumping membrane-bound protein involved in the respiratory pathway of many bacteria and is responsible for NAD$^+$ regeneration through NADH dehydrogenase activity (Marreiros et al., 2016). The structure of NDH-2 is relatively small and is approximately 42 kDa and binds non-covalently with a single flavin adenine dinucleotide (FAD) as the only coenzyme molecule (Kerscher, 2000). Type II NADH:quinoneoxido reductases (NDH-2) catalyse the two electron transfer from NADH to quinones, without any transfer of protons from cytosol to the Periplasm there by generating no energy (no proton motive force) at energy-transducing site. NDH-2 accomplish the turnover of NAD(P)H, regenerating the NAD(P)$^+$ pool, and may contribute to the generation of a membrane potential through complexes III and IV. There are a few compounds that can prevent their activity, but so far no general specific inhibitor has been assigned to these enzymes. However, they have the common feature of being resistant to the complex I classical inhibitors rotenone, capsaicin, and piericidin A. NDH-2 have particular relevance in yeasts like *Bacillus subtilis*, *Saccharomyces cerevisiae* and in several prokaryotes, whose respiratory chains are devoid of complex I, in which NDH-2 keep the [NADH]/[NAD$^+$] balance and are the main entry point of electrons into the respiratory chains.

1.6 Research problem statement

Microbial electrochemical systems (MESs) are emerging cost-effective, energy-efficient and environmentally sustainable technologies. Different applications of MESs can effectively utilise wastes and treat wastewater while concurrently recovering energy into various forms of energy like bio-electricity, bio-hydrogen and other value-added chemicals using the electrode setup. MESs generates potential which can be used drive thermodynamically uphill reactions like acetate production from CO_2, biohydrogen production using MEC, etc. Sustainability of MES lies in the use of self-regenerating biocatalyst which can use the organics directly from wastewater to generate bioelectricity with very less to no input of external energy. Performance of MESs depends on complex interactions between electroactive biofilms and electrochemical systems. Bacteria, acting alone as a pure culture or in syntrophically interacting biofilms can oxidise many different carbon substrates, making use of their versatile metabolic pathways. The performance of MES is tightly coupled with the efficiency of biocatalyst as bacterial metabolic pathways are linked to the electrochemical systems.

Despite having an edge over conventional wastewater treatment systems both in terms of environment and sustainability, microbial electrochemical systems still suffers from low power output. The main bottleneck for the MESs is low EET efficiency due to low electron flux which seriously restrained commercial applications of these MES systems. MES being a new generation of renewable energy technology perfectly couple microbial metabolism with electrochemical reactions. Highly efficient EET is essential for the industrial application of a diverse array of microbial electrochemical systems (MES) in environmental, energy fields and biorefinery domains. They have attracted growing interest from energy and biosynthesis industries and have been widely recognized as a promising technology for sustainable wastewater treatment, energy generation and biobased chemicals production. In addition to low electron flux, MES performance is constrained by several other factors such as poor cathode reaction kinetics, cathode reduction reaction, losses etc. In a recent metabolic modeling study, it was shown that not necessarily the degree of reduction of a product but rather the metabolic pathway that leads from the chosen substrate to the target compound decides if there is an electron surplus or demand inside the metabolic network of a cell (Frauke et al., 2015).

1.7 Scope of Research

For achieving sustainability, maximizing power output and enhancing resource recovery are highly desired for achieving sustainable systems. One important aspect which greatly affects the performance of microbial electrochemcial systems is the efficiency of electrogenic bacteria or community to transport electrons to the electrode through extracellular electron transport pathway. Poor electron transfer efficiencies of anodic bacteria or communities can result in high over potentials which will have negative impact on overall power output of MES/MFC. A better understanding of community structures of electroactive biofilms and extracellular electron transfer strategies may help to enhance the power density for practical application of MESs. One way to do this is by understanding how MES/MFC associated anodic or cathodic microbial communities are affected by environmental and process conditions. An appropriate method for selection of electrogenic or electroactive biomass is also essential for improving performance in terms of energy production and efficiency of the system. Selective enrichment will result in active electrogenic community which will allow for mutually beneficial syntrophic interactions involving the electrogens and acidogens and while suppressing nonproductive, energy-diverting bacteria like methanogens. An additional challenge associated with mixed culture inoculum is represented by population changes with time and loss of activity during storage. Advances in metagenomics, directed at obtaining knowledge on genomes of environmental microbes when combined with the conceptual framework of microbial ecology, will provide the tools to assess the ecological role of specific microbial phyla. This will enable us to establish strategies for the defined microbial ecosystems capable of producing stable bioelectricity sustainably and efficiently. Another strategy to enhance the power output of an MES is by improving the electron flux through extracellular electron transport using metabolic engineering strategies. Metabolic engineering involves regulating or redirecting metabolic networks by applying our knowledge of the cellular metabolic pathways. It usually involves manipulation of key enzyme levels either by overexpression or down-regulation to divert energy and resources to a specific pathway or for the production of required metabolite. In MES, metabolic engineering can be a powerful tool to increase/decrease the electron flux in the fuel cell.

Whether one is considering current consuming (eg: bioelectrochemical synthesis) or current producing (eg: microbial fuel cells) bio-processes, fundamental to each application is the performance of the biocatalyst or microorganisms. The ability and especially the efficiency of the biocatalyst to donate or accept electrons with an electrode and connect its extracellular electron transport to its cellular metabolism significantly affect the overall process output. Electron transfer rates in MESs are too low to design a viable scale-up process which has restricted them only to lab-scale projects (Logan and Rabaey, 2012). In order to optimize and advance microbial electrochemical systems for real field applications, a thorough understanding of possible extracellular electron transport mechanisms in both directions is essential (Tremblay and Zhang, 2015). Especially the fundamentals of electron transport towards microbes and cathodic systems are poorly understood (Rosenbaum et al., 2011; Sydow et al., 2014).

1.8 Objectives of the research

Based on the existing scope in this domain, the research was designed with the following specific objectives

- Genetically engineer *Escherichia coli* by over expression of *ndh* gene from *Bacillus subtilis*.
- Evaluation of the engineered strain for electrogenic and extracellular electron transfer activity by using electrochemical and biochemical assays.
- Enhancing extracellular electron flux by overexpression of NADH dehydrogenase II towards improving bioelectrogenic activity of MES.
- Develop strategies for development of electrogenic inoculum for MES application
- Understanding metagenomics using high throughput MiSeq platform to understand the relationship between selective enrichment and microbial composition.
- Evaluate the functional efficiency of microbial electrochemical system for power production and treatment of paper and pulp wastewater in defined conditions.

CHAPTER 2
Present State of Knowledge

2.1. Introduction

Microorganisms which have the ability to donate and accept electrons to and from anode and cathode are known as electroactive microorganisms or exoelectrogens and the process is called extracellular electron transport (EET). EET by electroactive microorganisms involve the flow of intracellular electrons derived from oxidative metabolism of carbon sources to extracellular electron acceptors (such as minerals and inert electrodes) via c-type cytochromes-based and/or soluble electron shuttle-mediated pathways. The application possibilities for electrode–bacteria interactions include the production of bioelectricity, waste and wastewater treatment, bioremediation and the production of valuable products therefore opening a wide research field (Fig 2.1) called microbial electrochemical systems which are environmentally sustainable and economically viable as they can use various low cost or waste substrate while producing value added compounds at no or very less energy inputs (Fig 2.3).

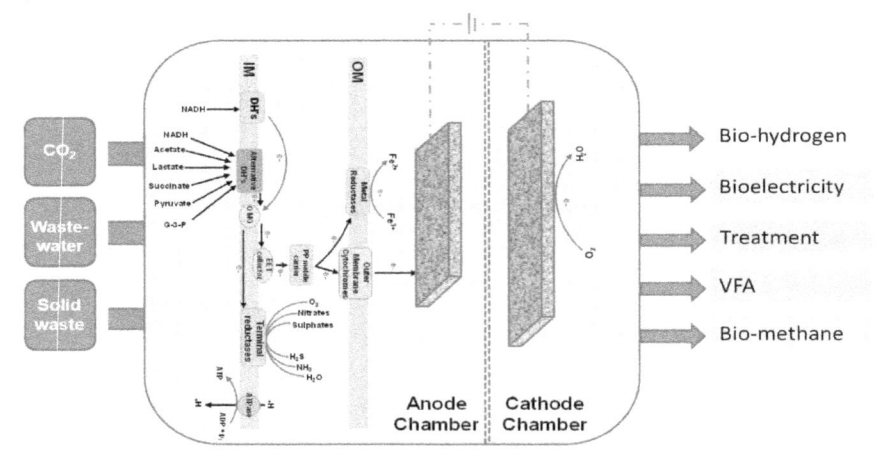

Fig2.1: Schematic diagram of MES and its applications.

2.1.1 Microbial Fuel cells (MFC)

Microbial fuel cell is the most studied type of MES application which can directly convert chemical energy present in an organic substrate to electricity through a cascade of redox reactions. The concept of producing bioelectricity using microbes was first conceived in 1911 by M.C. Potter where he used Saccharomyces to produce bioelectricity. MFCs use electrodes as electron acceptors (anode) and electron donors (cathode) to link up to the microbial metabolism. The potential difference induced by the introduction of electrodes will acts as a net driving force to transfer the electrons from anode to cathode. A conventional two-chamber MFC consists of anode (anaerobic) and cathode (aerobic) chambers separated by proton exchange membrane. In the anode chamber, electroactive microorganisms generate electrons through metabolism (catabolism) of organic compounds. Electrons travel to the cathode chamber through the external circuit, while protons conduct through the membrane where they are coupled to oxygen reduction which will result in production of water as final product which is clean. Electric current generated by MFC can be harvested by connecting an external resistor. Microbial cells were operated in several configurations, but single chambered MFC has become widely accepted because of the ease of operation and low ocst for its construction (Fig 2.2). Basic reactions occurring at anode and cathode chamber can be represented as below,

$$C_6H_{12}O_6 + 6\ H_2O \longrightarrow 6\ CO_2 + 24\ H^+ + 24\ e^-, (E^0 = 0.014V) \qquad \text{Eq 2.1}$$
$$6\ O_2 + 24\ H^+ + 24e^- \longrightarrow 12\ H_2O, (E^0 = 1.23\ V) \qquad \text{Eq 2.2}$$

MFCs can utilize a wide range of organic biomass as anode substrate (anolyte) and can be integrated with the existing effluent treatment units increasing its economic viability and sustainability.

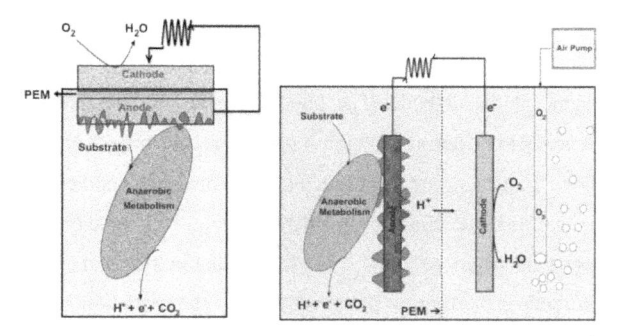

Fig 2.2: Schematic diagrams of single and dual chambered microbial fuel cells

2.1.2 Bioelectrochemical synthesis (BES)

Bioelectrochemical synthesis (BES) refers to the production of higher carbon compounds by bacteria in a BES under current consumption. The term BES was initially used only for the microbial reduction of CO_2 with the help of electricity to higher carbon products (Fig 2.3). As time progressed, the scope of BES has expanded to other carbon sources in addition to carbon dioxide which is also often referred to as electrofermentation. In nature, homoacetogenic bacteria reduce carbon dioxide to acetic acid and other higher carbon chemical with hydrogen as the reducing agent or electron donor. However, supplying acetogens with hydrogen is impractical because of the cost it would add to the process. An alternative to this problem is to supply the electrons (reducing equivalents) to the acetogens directly using electrodes. Main principle of BES is to overcome the metabolic redox limitations by using electron supplied by external power source (via cathode) for increased production of biobased products. All electrons needed for the half-reaction at the cathode are generated by the anoxic oxidation of wastewater at the anode. But to meet thermodynamic requirements for synthesis of product, additional electrons are needs to be supplied to the system using potentiostat or external power source. The electro-microbial synthesis of chemicals is one of the most potential applications for microbial electrochemical technologies as the process is sustainable and environmental friendly when compared to the non-renewable chemical synthesis. Carbon capture process of MES technology can significantly contribute to lower CO_2 emission which is one of the big driving forces for MES advancement and therefore have a positive effect on climate change and global warming.

2.1.3 Microbial electrolysis cell (MEC)

The low current densities of MFCs have encouraged a variation of microbial electrochemical system that couples the anodic decomposition of organic wastes with the production of hydrogen or methane at the cathode. The key difference of MEC to an MFC is that an additional voltage needs to be applied, which supplements the low potentials generated on the anode sufficiently to reduce protons, producing biohydrogen. Another difference between MFC and MEC is that, cathode chamber in MES is maintained at anaerobic condition which prevents spontaneous electricity generation in MEC unlike in MFC. In MEC, all the electrons required for the cathode half reaction are generated by the oxidation of organic susbstrates present in wastewater at the anode. As part of the energy for this reduction reaction is derived from the bacterial metabolic activity, the total energy that has to be supplied to the systems is less (0.2–0.8 V) than for electrochemcial electrolysis (1.23–1.8 V) of water. MECs have wider substrate range and high H_2 production when compared to the traditional fermentative H_2 production from wastewater. Typical reaction of hydrogen production in MEC using acetate as carbon can be represented as,

$$\text{Anode chamber:} \quad C_2H_4O_2 + 2\ H_2O \longrightarrow 2\ CO_2 + 8\ e^- + 8\ H^+$$
$$\text{Cathode chamber:}\ 8H^+ + 8e^- \longrightarrow 4\ H_2$$

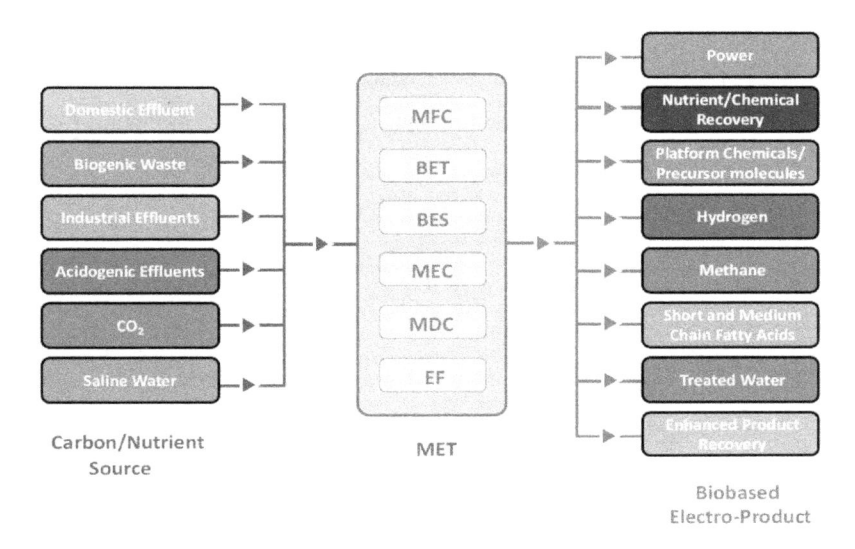

Fig 2.3: Applications and scope of microbial electrochemical system and the wide range of substrates they can use as substrates.

2.2 Biocatalyst in MES

Microorganisms exhibiting electrocatalytic activity have the potential to generate reducing equivalents (protons and electrons) through their metabolic reactions and transport them to an external electron acceptor (Logan and Rabaey, 2012). This electrocatalytic activity of a biocatalyst is the basis for the development of microbial electrochemical technologies (METs). Anode and cathode compartments separated by an ion exchange membrane provide a perfect environment for harnessing reducing powers generated by the electroactive microorganisms into potential energy (Butti et al., 2016). This can be converted into various usable forms such as bioelectricity in the case microbial fuel cells (MFCs) and bio-based products in the case of a bioelectrochemical synthesis system (BES). Moreover, these high energy electrons (in the form of reduced coenzymes such as NADH, FADH$_2$) can also be used for the breakdown of pollutants in bioelectrochemical treatment systems. In nature, a great diversity of electroactive microorganisms have been identified belonging to eubacteria (Liu et al., 2002), archae (Dopson and Holmes, 2014), fungi (Fernández de Dios et al., 2013), algae (Chandra et al., 2012), photobacteria (Venkidusamy and Megharaj, 2016), and cyanobacteria (Lea-Smith et al., 2016). These electroactive microorganisms are often isolated from wastewater, lake sediments, soils, etc. and then subjected to selection in the electrogenic environment, such as in MESs (Fig 2.4). Electroactive microbes isolated from diverse sources tend to exhibit a wide range of substrate utilization capability as they can oxidize an extensive range of biodegradable compounds including fatty acids, carbohydrates, alcohols, gases such as CO_2, and even relatively recalcitrant materials such as cellulose (Venkata Mohan et al., 2014b).

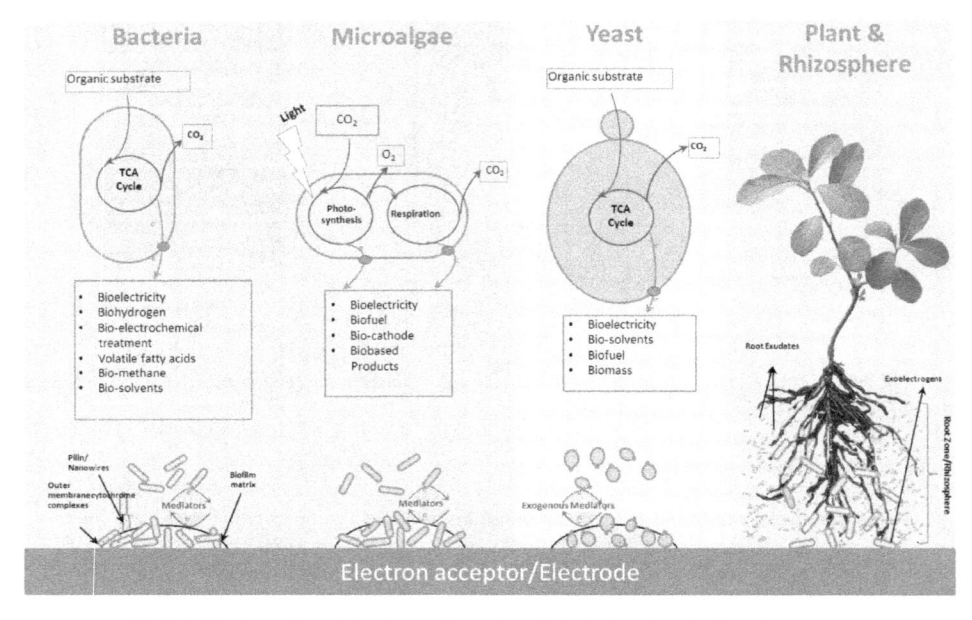

Fig2.4: Various biocatalysts used in microbial electrochemical systems.

In MES, both pure and mixed cultures were used as inoculum and both have their respective advantages and disadvantages. Pure cultures are more beneficial for investigating the mechanism of extracellular electron transportation in electroactive microbes and for production of specific chemicals in bioelectrochemical synthesis system. But the pure culture biofilms are more susceptible to environmental changes and are less effective for real field applications like wastewater treatment. More over the maintaining the purity of the culture needs extra inputs of energy. Unlike pure culture mixed culture biofilms are more robust and resilient to the environmental stress and provides multiple environmental niches which are often advantageous when using wastewater as the substrate.

Microbes interacting at anode and cathode interface to function as electrode-reducing and electrode-oxidizing microorganisms, respectively (Semenec and Franks, 2014). However, a major challenge in the development of efficient MET is to understand the link between electroactive microbes and electrodes. This can be done by understanding the fundamental chemistry of biocatalyst attachment, interaction with electrodes, and charge transport at electrode. Moreover, solving the molecular structures of the surface interacting proteins associated with electron transfer at biofilm electrode interface will help in fine-tuning the activity of MESs.

2.3 Types of biocatalyst used in MESs
2.3.1 Bacteria

Many of the bacteria are generally known to grow aerobically using oxygen as a terminal electron acceptor (TEA) to generate ATP, whereas others are known to grow at low concentrations of oxygen (anoxic) or in the absence of oxygen (anaerobic). Unlike aerobic bacteria, anaerobes can use various kinds of TEA-like inorganic compounds such as nitrates, sulfates, iron, DMSO manganese, etc. Bacteria which gain energy by donating electrons to metal oxides (ores) resulting in the reduction of these metal ions are termed as metal-reducing bacteria (Venkata Mohan et al., 2014a). These bacteria help in a redox cycling of metals, immobilization of radionuclides, and degradation of contaminants in various environments such as lakes, sediments, aquifers, etc. (Ilbert and Bonnefoy, 2013). They have evolved strategies to transfer the electrons directly to the external electrode or metal ore during their metabolism, which is difficult considering that the cellular membrane is resistant to transfer of charged molecules due to its non-polar (lipid) nature. This extracellular electron transfer (EET) ability of bacteria has shown to be involved in various environmental phenomena such as biogeochemical cycles, pollutants removal, etc. Geobacter sulfurreducens and Shewanella oneidensis MR-1 are considered as the model organisms for studying the characteristics of electroactive bacteria (Fig. 1.6.1). In addition to oxygen and nitrates, these two bacteria are also known to use iron, manganese, sulfur, and nickel ions as TEAs. Despite identification of many electroactive bacteria (EAB) by high-throughput sequencing analysis, only a small number of EAB were isolated and studied for their EET mechanisms. Specific growth requirements and functional gene markers for identification of the EAB and tools to isolate them are under investigation (Ilbert and Bonnefoy, 2013).

2.3.1.1 Electrode-Microbe interactions

Electroactive bacteria have developed specialized molecular conduits to transport electrons across cell membranes, enabling these bacteria to utilize extracellular electron acceptors and donors during respiration (Lovley, 2012; Shi et al., 2016). Transport of electrons from the cell surface to extracellular electron acceptors and from donors is most complicated and depends upon the redox potentials of electron donor and acceptors (Fig 2.5). There are various challenges that bacteria face at microbe-mineral or microbe-electrode interfaces.

standard reduction potential $E^{\prime 0}$ (mV)

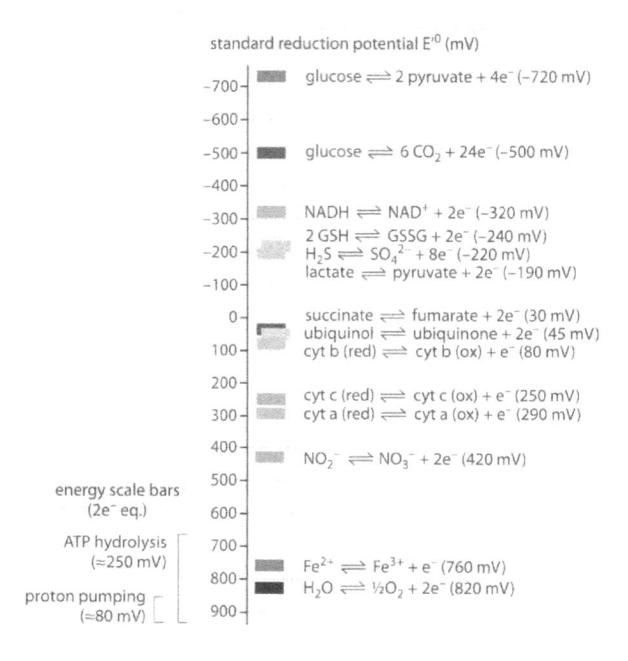

- −700: glucose ⇌ 2 pyruvate + 4e⁻ (−720 mV)
- −500: glucose ⇌ 6 CO_2 + 24e⁻ (−500 mV)
- −300: NADH ⇌ NAD^+ + 2e⁻ (−320 mV)
- −200: 2 GSH ⇌ GSSG + 2e⁻ (−240 mV); H_2S ⇌ SO_4^{2-} + 8e⁻ (−220 mV)
- −100: lactate ⇌ pyruvate + 2e⁻ (−190 mV)
- 0: succinate ⇌ fumarate + 2e⁻ (30 mV)
- 100: ubiquinol ⇌ ubiquinone + 2e⁻ (45 mV); cyt b (red) ⇌ cyt b (ox) + e⁻ (80 mV)
- 200/300: cyt c (red) ⇌ cyt c (ox) + e⁻ (250 mV); cyt a (red) ⇌ cyt a (ox) + e⁻ (290 mV)
- 400: NO_2^- ⇌ NO_3^- + 2e⁻ (420 mV)
- 800: Fe^{2+} ⇌ Fe^{3+} + e⁻ (760 mV); H_2O ⇌ ½O_2 + 2e⁻ (820 mV)

energy scale bars (2e⁻ eq.)
ATP hydrolysis (≈250 mV)
proton pumping (≈80 mV)

Fig 2.5: Standard redox potentials of various redox proteins and molecules of electron transport chain interacting with electrode (Standard redox potentials are calculated at pH 7 and 25°C Vs SHE)(Milo and Phillips., 2015).

The transformation of iron or other metals between different mineral states is one of the examples. For example, oxidation of ferric and ferrous forms by iron-oxidizing acidophilic bacteria is not a major challenge as both are available in soluble forms. While in the case of iron-oxidizing neutrophilic bacteria, the challenge is substantial as the bacteria have to prevent the accumulation of insoluble forms of irons on its surface. Bacteria living in suspension can accept electrons from the surface of insoluble metal forms or electrodes by interacting with metal substrate briefly (Bond et al., 2012). Whereas the bacteria living in the biofilm of an electrode or mineral ore face challenge depending on location of bacteria in biofilm. Bacteria located at the surface of an electrode or mineral ore have the highest availability of terminal electron acceptor while having the lowest availability of nutrients because of the several layers of biofilm hindering the diffusion of nutrients to the electrode surface. While in case of bacteria living on the surface of biofilm, which high concentrations of nutrients but are in less contact with electrode surface due to the biofilm present in between electrode and bacteria.

Biofilm housed bacteria living on the surface have several ways to transport electrons from the biofilm surface to the electrode surface, which can be differentiated into direct, indirect, and mediated mechanisms (Fig 2.6). Direct electron transfer (DET) mechanisms involve direct contact between the bacteria and the electron acceptor such that electron transfer proteins and various cytochromes on the outer membrane surface could make contact and transfer electrons onto the electrode directly (Lovley, 2012). Indirect electron transfer involves the utilization of extracellular wire-like appendages such as pili that allow electron flux from the electroactive bacteria to an electron acceptor. These extracellular wire-like appendages are also known as nanowires and can conduct electricity at several cell length distances (Kracke et al., 2015a). In mediated electron transfer, soluble redox mediators either from cellular metabolism or external sources are utilized to shuttle electrons between the electrode and bacteria. Most of the mediators can move across the cellular membrane, making them ideal for shuttling electrons between electroactive bacteria and the electrode. Mediators such as flavins, phenazines, pyocyanins, and siderophores are from the cellular metabolism, whereas humic substances, chemical mediators, and environmental mediators can act as external mediators. All the three mechanisms contribute and function under different environmental stresses (Richardson et al., 2012a; White et al., 2016).

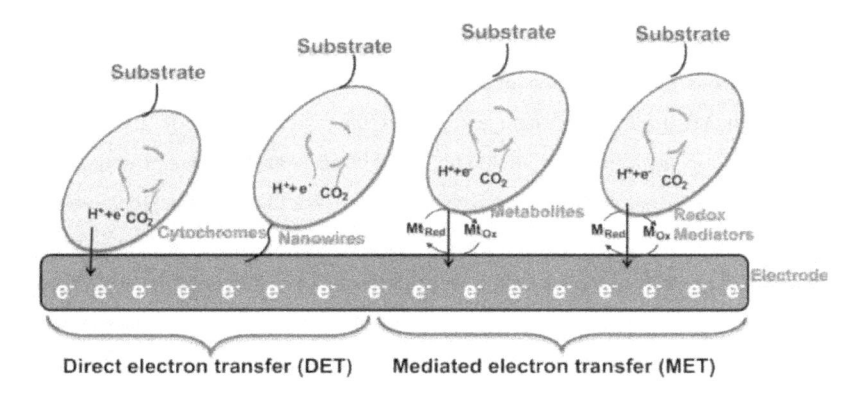

Fig 2.6: Modes of extracellular electron transport in microbial electrochemical systems a) direct b) nanowires and c) mediators.

2.3.1.2 Electron transport across cell membrane

The majority of organisms that perform the metabolic oxidation or reduction of metal species transports electrons to the cell surface where the catalytic redox reaction takes place. Bacteria that utilize minerals and electrodes as TEAs require the transport of electrons generated in the cytoplasm to the surface of the cell (Lovley, 2013; Marsili et al., 2008). The outer membrane of gram-negative bacteria is an insulating barrier that requires a conduit for electron passage, whereas in gram positive bacteria, the cell's outer wall acts as a barrier between the cell and the mineral surface. Consequently, in order to transfer electrons from inside the cell to the cell surface and vice versa, bacteria use a diverse range of cofactor-rich proteins (Sydow et al., 2014a). A wide range of microbes have been discovered to be able to exchange electrons with solid surfaces (direct EET) and/or via mediators (indirect EET). The diversity of mechanisms of electron transport is so high that no two electroactive bacteria identified till now have same mechanism (Kracke et al., 2015b). Dissimilatory metal-reducing bacteria are among the well-studied EAB able to respire insoluble metals in anaerobic environments (Lovley, 2013). The model organisms *Shewanella oneidensis* and *Geobacter sulfurreducens* were studied in detail by several research groups and evidences were found for direct and indirect electron transfer (Lovley, 2012). The current understanding of the EET mechanisms for transport of electrons across the cell envelope, between metal ores and electrodes, is under study.

2.3.1.3 Extracellular electron transport- gram negative bacteria

It has been found that the gram-negative bacteria use two types of electron transport systems by which electrons get transported inside and outside of the outer membrane. The first system, the MtrAB porin cytochrome complex is the representative of many complexes identified from the metal/electrode-reducing bacteria such as *Shewanella* and *Geobacter* (White et al., 2016). The second transport system is the Cyc2 fused porin-cytochrome system that is commonly used by the iron-oxidizing bacteria when extracellular iron is present in soluble form. The best example for such a system is found in *A. ferrooxidans*, which grows at acidic conditions where iron is only present in soluble form (Liu et al., 2014). Two EET pathways are well studied in case of gram negative bacteria one of which is MTR pathway of *Shewanella oneidensis* and the other one is nanowires of *Geobacter sulfurreducens*. MTR pathway is the representative of MtrAB porin-cytochrome system.

2.3.1.4 MTR pathway in *Shewanella*

The metal-reducing or Mtr pathway of *S. oneidensis MR-1* was one of the best studied extracellular electron transport pathways, and it can use various minerals that contain Fe (III), Mn (III), or Mn (IV) as TEAs (metal reduction) (Shi et al., 2012). Studies have shown that this pathway comprises six multiheme c-type cytochromes CymA, FccA, MtrA, OmcA, MtrB, MtrC, and small tetraheme cytochrome (STC) and the porin-like MtrB, which is an outer membrane protein. CymA is involved in the oxidation of quinol from the quinone pool in the cytoplasmic membrane and transfers them to the periplasmicFccA and STC. MtrA, MtrB, and MtrC help in the transfer of electrons from the CymA to the outer membrane protein complex that transfers electrons to insoluble electron acceptors (Coursolle et al., 2010). MtrA, MtrB, and MtrC form a trans-outer membrane protein complex MtrCAB where MtrC a decaheme cytochrome and OmcA a lipoprotein are present in the outer membrane. These two proteins can physically interact and transfer electrons extracellularly. MtrB is a porin complex present in the outer membrane and also serves to stabilize MtrA and MtrC by anchoring them. MtrC and OmcA were also found to be associated with extracellular structures such as nanowires. In addition to the DET, *S. oneidensis* MR-1 also transports electrons extracellularly by releasing redox mediators called flavins (Kracke et al., 2015b). Gene *bfe* was found to control the extracellular release of flavins. In addition to the anode side, Mtr pathway is also known to be involved in cathode reactions but the information about the electron transport in the cathode (electrode to bacteria) chamber is limited (Ross et al., 2011).

2.3.1.5 Nanowires- *Geobacter*

Geobacter sulfurreducens is a gram-negative obligate anaerobe belonging to class *δ-proteobacteria*. It is considered a model organism for investigating extracellular electron transport and electroactive microorganisms (Bond et al., 2012). Its genome has more than 110 genes coding for putative c-type cytochromes, which are known to play a key role in the extracellular electron transport pathway of this bacterium (Kracke et al., 2015b). In *G. sulfurreducens* PCA, multiheme c-type cytochromes have a key role in EET across the cell envelope during the extracellular reduction of electrodes and minerals (Fig. 1.6.1). These cytochromes enable the bacteria in channeling the electrons from the menaquinone pool to the extracellular insoluble metals and electrodes (Shi et al., 2016). Various multiheme c-type cytochromes identified in *Geobacter* are the putative quinol oxidases ImcH and CbcL located in the cytoplasmic membrane, PpcA and PpcD present in the periplasm which transfers electrons to the various OMCs (e.g., OmcB, OmcC, OmcS, and OmcZ). OMCs are actually the ones which interact with the external electron acceptors and donate electrons to electrode or mineral ore. It has also been found that OmcZ is abundant in the biofilms grown on the electrodes, whereas OmcB and OmcS are known to be important for reduction of iron minerals but have very less effect on current production (Richter et al., 2012). *G. sulfurreducens* shows highest electroactivity as biofilm grown on the electrode. These microbial species possesses excellent biofilm formation ability and extracellular electron transport is mostly DET (Kracke et al., 2015b). *Geobacter* biofilm thickness is linked with the amount of current (upto 5 mA) generated in a linear correlation up to a certain thickness (40e50 mm). They depend on nanowires, which are type IV pili-like structures and are known to be electrically conductive. Nanowires enable them to have physical connection between bacteria and electrode surfaces. The mechanism of the electron transport in nanowires is still under study, with one model proposing that nanowires possess metallic-like conductivity which relies on several aromatic amino acid residues located on the nanowires which enable electron delocalization on nanowires due to p-stacking (Malvankar et al., 2015), while another model proposes that the electrons on nanowires are transported by electron hopping along the chain of various redox proteins which is known as the super-exchange model (Bond et al., 2012).

2.3.1.6 Extracellular electron transfer- gram positive bacteria

Despite many reports on identification of gram-positive bacteria in electrogenic consortiums, no defined EET mechanism has been identified till now (Carlson et al., 2012). As the outer membrane is absent in gram-positive bacteria, it was assumed that there is no requirement for EET systems such as porin cytochrome systems unlike gram negative bacteria. However, presence of thicker cell walls may limit the ability of these bacteria to transfer electrons extracellularly without extracellular electron transport system (Richardson et al., 2012b). Several gram-positive bacteria were identified in current-producing anode biofilms such as *Thermincolapotens strain* JR (Wrighton et al., 2008), which achieved a columbic of 91% when acetate was used as the substrate. Imaging studies confirmed that the *T. potens* use direct EET to transport electrons to the electrode. Genome studies identified several c-type cytochrome coding genes which are the epitome of extracellular electron transport. *Carboxydothermus ferrireducens*, another gram-positive thermophilic bacterium known for its ability to reduce insoluble iron forms, was shown to utilize DET (Gavrilov et al., 2012).

The EET system used by C. ferrireducens is yet to be understood. Initial studies have shown that it contains cytochromes and conductive pili when grown in presence of insoluble electron acceptors. One such cytochorme is, enzyme cytochrome Fe-(EDTA) reductase which has shown the ability to reduce the iron forms (Wrighton et al., 2008).

2.3.2 Yeast

Studies have shown that eukaryotic organisms such as yeast can also be used in MET as they are easy to handle, nonpathogenic, capable to utilize various substrates for growth, and can grow in diverse aerobic and anaerobic environmental conditions (Haslett et al., 2011). Several yeast species such as *Arxulaadeninivorans* (Haslett et al., 2011), *Saccharomyces cervisiae* (Kumar et al., 2016), *Hansenulaanomala* (Prasad et al., 2007), and *Candida melibiosica* (Hubenova and Mitov, 2010) were studied in METs. Experiments with yeast biocatalyst in MFC for the generation of electricity resulted in lower power outputs than bacterial cells, which were initially thought to be due to the low catabolic rates of yeast. Studies with the use of electron transfer mediators such as methylene blue, methyl orange etc., has shown increased power output at similar growth rates. This suggests that the catabolic rates in yeast and mesophilic bacteria are similar but the low power outputs was due to difficulty in accessing the intracellular electron transfer mediators (Gunawardena et al., 2008). The studies have also shown that electrode modifications also increase the power production in yeast MFC (Hubenova and Mitov, 2010). The electron transport mechanisms in yeast depends on the respiration, glycolysis, and the trans-plasma membrane electron transport proteins which takes place inside the cell enclosed by the cell wall and plasma membrane (Fig 2.7). EET in yeast can be of direct mode through membrane-bound cytochrome-based proteins or mediated by using exogenous electron transfer mediators (Fig. 1.6.3). *Candida melibiosica* was shown to transfer the electrons by combining two mechanisms involving the expression of cellular endogenously generated mediator(s) and by the formation of yeast biofilm on the electrode surface (Hubenova and Mitov, 2010). In some species such as *A. adeninivorans*, the electron transfer is through secretion of redox molecules as well as through DET, whereas in S. cerevisiae, redox mediators are not produced indigenously, and hence use of this organism in fuel cells requires exogenous mediators for EET. The DET from the yeast to the anode of the MFC was also reported in *S. cerevisiae* (Prasad et al., 2007). MFC operated with *Candida melibiosica* as a biocatalyst in the anode region and potassium ferricyanide as an electron acceptor in the cathode region resulted in good power output (Hubenova and Mitov, 2010).

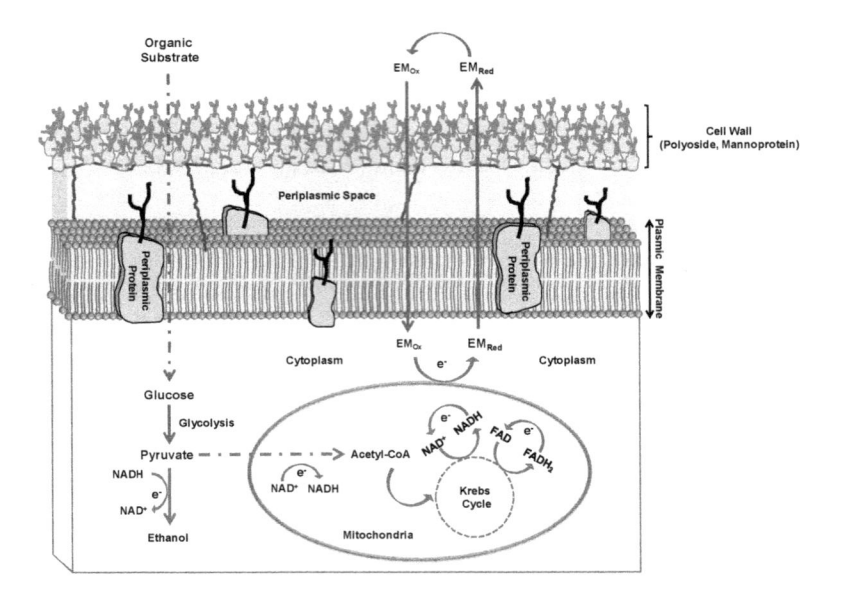

Fig 2.7: Role of biocatalyst at bioanode: Extracellular electron transport in yeast cells. In yeast cell, electrons are transported to electrode by soluble redox mediators which can be metabolically generated or from externally source.

Another report using carbon felt electrode with surface nickel nanostructures resulted in significant increase in power density, which shows that electrode modification enhance power output in yeast fuel cells (Raghavulu et al., 2011). *S. cerevisiae* was also used as biocatalyst in MFC and was shown to produce highest power at pH 6 (Gandasasmita, 2016). Another strain of yeast which was also successfully used as a biocatalyst in MFC was *Arxula adeninivorans* (dimorphic yeast) (Haslett et al., 2011). It is a robust organism which can grow at high temperatures (upto 48°C), tolerate pH range of 2 to 10, and high salt concentrations (upto 20% NaCl) and can use potassium permanganate as the electron acceptor in the cathode region (Haslett et al., 2011).

2.3.3 Fungi

Fungi are members of the eukaryotic group of microorganisms. The structure of fungi contains long, branching filamentous structures known as hyphae, which are primary structures of vegetative growth in most of the fungus. As fungi do not exhibit EET, these organisms cannot act as a direct biocatalyst in MET. However, these multicellular organisms can act as intermediary structures between a biocatalyst and electrode with the aid of their

Chapter 2 Present State of Knowledge

hyphal structure (Fernández de Dios et al., 2013). Fungi can also be used in MET for remediation of wastewater containing hazardous chemicals. In addition to this, fungus metabolic activity under varied environmental conditions can produce intra- and extracellular enzymes. These enzymes help fungi survive in adverse environmental conditions, making them a good option as co-biocatalyst in MET applications.

2.3.4 Algae

2.3.4.1 Algal bioanode

Algae-based METs function with minimum energy, and the ability of algae to form syntrophic association with bacteria will lead to higher substrate degradation efficiency and power output (Lea-Smith et al., 2016). Microalgae are eukaryotic cells containing mitochondria which is initially thought to be the main source of extracellular electrons (Li et al., 2015). Algae when co-cultured with photobacteria as a biocatalyst in the anode chamber showed to have light-dependent electroactivity. When exposed to sunlight they showed increased power output compared to dark condition, indicating the involvement of photosynthetic machinery in EET (Lea-Smith et al., 2016; Pisciotta et al., 2011) and wastewater treatment (Gajda et al., 2013).

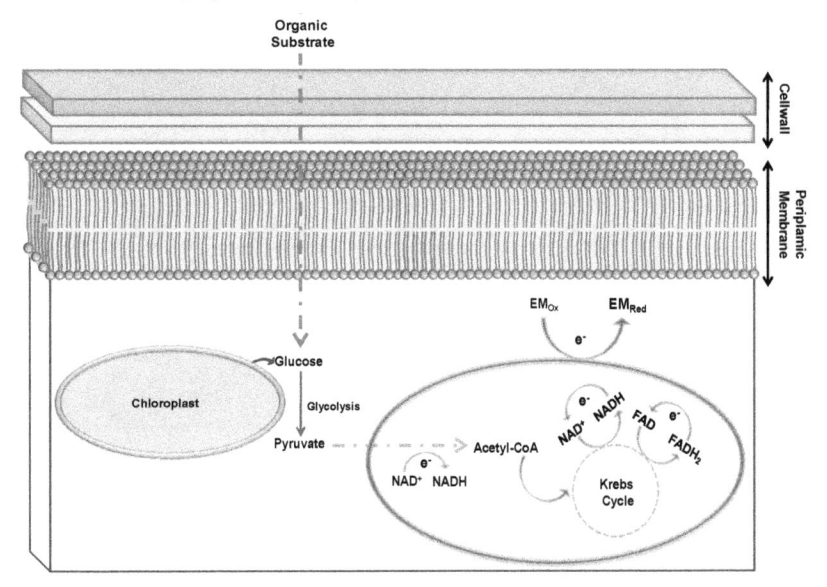

Fig 2.8: Role of algal biocatalyst at bioanode and respective extracellular electron transport.

Chattonella marina, a marine algae can perform light-induced electroactivity independently (Pisciotta et al., 2011), and its exudates contain various redox active compounds such as redox active proteins like cytochrome-based proteins, quinones, and redox mediators. The source of electrons in algal bioanode is both mitochondrial and photosynthetic electron transport chains (Fig. 2.8). Several phototrophic cyanobacteria such as Lyngbya, Nostoc, and several microalgal species of *Chlorella* were reported in MFC operations(Liu and Choi, 2017).

2.3.4.2 Algal biocathode

METs have relied on precious metals and artificial redox mediators for reduction of oxygen at the cathode. Algae at cathodes (algal biocathode) for power generation can be an effective alternative for the metal based oxygen reduction reaction (Fig 2.9). The function of algae at the cathode was found to be different in the presence or absence of sunlight (Chisti, 2007).

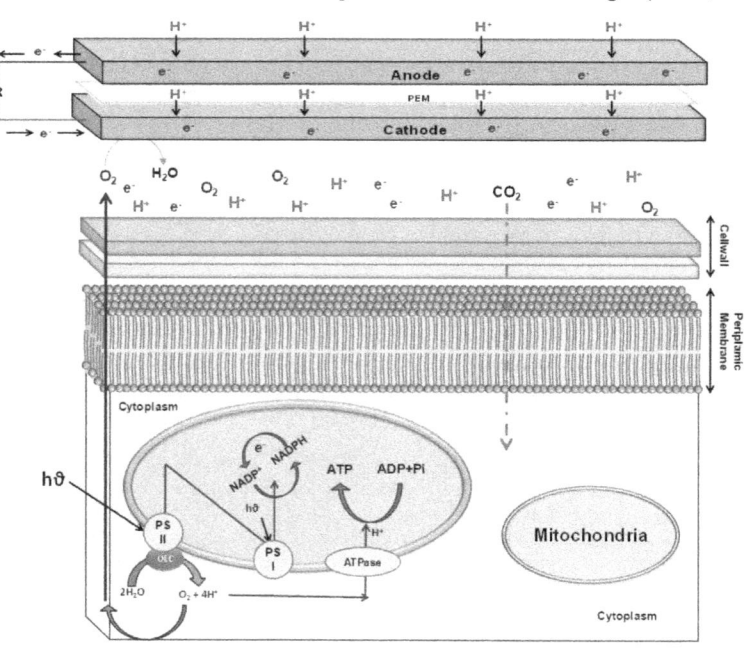

Fig 2.9: Role of biocatalyst in biocathode: Microalgae at biocathode. Oxygen produced by the microalgae will accept the electrons as cathode enhancing the terminal step.

In the presence of sunlight, algae are known to perform photosynthesis where light energy is converted to chemical energy, i.e., conversion of CO_2 to organic matter and oxygen in the presence of sunlight. The biomass accumulated during this process can be used to produce energy and other value added products (Saba et al., 2017). Algae were also used as biocatalyst in other MET applications like microbial desalination cell where *Nannochloropsis salina* was used for salts removal at cathode (Tharali et al., 2016) and H_2O_2 producion by *Chattonella marina* at the cathode (Pisciotta et al., 2011).

2.3.5 Photosynthetic bacteria

The scope of METs for sustainable power production has increased substantially in recent years with the identification of diverse bacterial populations that associate syntrophically with the electroactive microbes for a constant energy supply. In this regard, photosynthetic bacteria were considered as a good candidate as they harvest solar energy and link its metabolic activity with EET. The use of photosynthetic bacteria in MFC for production of bioelectricity will also help to reduce the carbon dioxide levels in the atmosphere due to coupling of photosynthesis with bioelectricity generation (He et al., 2009). In addition to functioning as pure biocatalyst, photosynthetic bacteria is also known to form synergistic interaction with heterotrophic bacteria which resulted in high power production without addition of external carbon source (He et al., 2009). Photosynthetic bacteria in bioelectrochemical systems have shown high dechlorination and demineralization of 4-chlorophenol with simultaneous power production (Kong et al., 2014). *Rhodopseudomonas palustris* DX-1 is the most studied photosynthetic bacteria which showed high power densities when used as an anodic biocatalyst (Xing et al., 2008).

2.3.6 Cyanobacteria

Cyanobacteria as a biocatalyst in MET have shown some interesting developments in bioelectricity generation by utilizing solar energy. Like in case of algae, cyanobacterial photocurrent is also light dependent, which increases with the increase in light duration. These photosynthetic organisms can act as an efficient biocatalyst in both anode and cathode region by making use of its photosynthetic mechanism. High power output was reported by cyanobacterial species (*Synechocystis*) in microfluidic cell (Bombelli et al., 2015). Immobilized Nostoc sp. coated on the surface of the carbon nanotubes (CNTs) of anode has shown light-dependent power production at +0.2 V where immobilization has shown to enhance direct electron transport to anode (Sekar et al., 2014). The study has showed that plastoquinone pool and quinone oxidases are involved in the transport of electrons from PS-II to CNT. Microcystis aeruginosa as a biocatalyst in the cathode have shown that reactive oxygen species released by cyanobacteria helped in the production of electricity (Cai et al., 2013). A study on diverse genera of cyanobacteria has reported some interesting findings on the electrogenic activity (photosynthetic electron transfer chain) and concluded that the mechanism of electron transfer in cyanobacteria appears to be principally different from the electroactive organisms identified earlier (Pisciotta et al., 2010). *Synechocystis* is known to produce type IV pili, which is shown to have high conductivity and can only be produced under CO_2 limitation and excess light conditions (Gorby et al., 2006). Biophotovoltaic cells using cyanobacteria coated on the CNTs were reported to produce current continuously in light and dark conditions (Sawa et al., 2017).

2.5 Quorum sensing

Quorum sensing is one of the many signaling mechanisms involving cell-cell communication. It is widespread among the bacteria and serves as a global regulator for a variety of bacterial genes at a cell-density dependent manner (Gimkiewicz and Harnisch, 2013). By employing quorum sensing, bacterial communities can coordinate individual bacterial behavior into wide range of social behaviors like pathogenesis, biofilm formation etc. Quorum sensing and its role in BES were extensively studied in *Pseudomonas* which is a gram negative bacterium. It primarily produces bioelectricity by secretion of redox mediators into the anolyte which shuttle electrons between bacteria and electrode. Redox mediators secreted by *Pseudomonas* primarily comprise of phenazines which are heterocyclic nitrogen containing secondary metabolites which can act as redox shuttlers. Production of phenazines is regulated by various quorum sensing systems like *rhl, las,* and PQS which are organized in a complex regulatory network. Its synthesis is mainly regulated by *rhl*/QS system. In addition to these PqsE system also up-regulate phenazines synthesis in *Pseudomonas* independent of PQS system by up-regulating the *rhl* system. It was observed that the when rhlI and rhlR genes are overexpressed, it resulted in up-regulation of rhlA gene resulting in higher bioelectricity due to the overproduction of phenazines (Torres et al., 2009). It has been found that the overexpression of PqsE gene which is part of PqsABCDE operon controlling phenazines production and independent of PQS signaling system and oxygen can result in enhancement of bioelectricity even in the absence of oxygen. This is advantageous as anode chamber in BES in anaerobic and PQS systems responsible for phenazines production requires oxygen to express, whereas PqsE does not and can independently produce without the oxygen in BES.

In addition to this quorum sensing is also controlled by several global regulators. One such example is GacS/GacA which is a two-component response regulator and up-regulates the *las* and *rhl* systems. When RetS which is a sensory kinase interacts with GacS, it negatively regulates the phenazines production and bioelectricity. While the GacA overexpression leads to the production of phenazines results in bioelectricity production. It has been shown that the *retS* knocked out strain can deliver up to 45 times higher current than wildtype (White et al., 2016). Quorum sensing was also known to regulate synthesis of biofilm by controlling the production of extracellular polymeric substance (EPS), which was shown to reduce the conductivity of biofilms in MESs.

2.6 Mixed culture biocatalyst for MES application

The output of MET in the form of bioelectricity or any other bio-based product is due to the electrogenic activity of the biocatalyst used in the study. In MET, biocatalyst can be used in either pure or mixed form depending on the output of the study. The electrogenic activity of biocatalyst is influenced by many factors such as organic load of the substrate, electron transfer rate, mode of operation and other physico-chemical properties. These microbes delivers electron to their external environment by metabolically oxidizing the substrate provided thus showing the electroactivity which is a fundamental of microbial electrochemical technologies (Lim et al., 2017). Both pure and mixed cultures can be used as biocatalyst in BES, but due to high operational costs, pure cultures are not suggested for MES operation as the cost for maintenance towards purity of culture is not economic and the current densities generated are also low compared to mixed culture based BES. Hence, mixed cultures are more widely used in METs as they are cost effective and have several advantages over pure cultures. Usage of mixed inoculum in MESs means cost of the process is reduced drastically which is advantageous when they are used in wastewater treatment. Syntrophy in the mixed biocatalyst helps in degradation of wide range of complex substrates leading to effective removal of pollutants and COD when wastewater is used as anolyte. However the mixed culture contains significant amount of non-electrogenic (acetogenic, fermentative, methanogenic and denitrifying bacteria) bacteria which consume organic substrate without generating electricity. This limitation in mixed biocatalyst can be eliminated by pretreating the mixed culture before reactor startup i.e. inoculating the reactor.

CHAPTER 3

Enhancing Extracellular Electron Flux: NADH Dehydrogenase II Overexpression to Improve Bio-electrogenic Activity

3.1 Introduction

MES have made possibility of treating organic pollutants in wastewaters with concurrent generation of energy in the form of bioelectricity. They use electroactive bacteria (EAB) as biocatalysts and their electrode interactions to drive electrons from oxidation of organic compounds for bioelectricity generation (Logan, 2009; Santoro et al., 2017; Venkata Mohan et al., 2014b). To date a wide range of EAB were identified that exchange electrons with insoluble electron acceptors either directly or indirectly (Kracke et al., 2015b; Lovley et al., 2011; Sydow et al., 2014b). Indeed, the electron transfer mechanism of each species varies significantly from each other such that there are several mechanisms for direct electron transfer, mediator electron transport and other modes of extracellular electron transport depending on the organism. However, there are several bottlenecks in up scaling of MET systems; especially due to the low rate of extracellular electron transfer. Recent years has seen several studies focused on genetic engineering approaches to optimize the rate of extracellular electron transfer and enhancement of electroactive biofilm formation to obtain high power yields (Han et al., 2016; Jensen et al., 2010).

Engineering efficient, directional electronic communication between living and nonliving systems has the potential to combine the unique characteristics of both materials for advanced biotechnological applications. However, the cell membrane is designed by nature to be an insulator, restricting the flow of charged species; therefore, introducing a biocompatible pathway for transferring electrons across the membrane without disrupting the cell is a significant challenge. However, low EET efficiency remained a crucial bottleneck that seriously restrained commercial applications of these BES systems. Reduced form of NAD^+ (NADH) acts as the major reserve for intracellular electron pool and the source of electrons and protons for EET. It is thus conceivable that manipulating intracellular reducing equivalents through increasing NADH regeneration may lead to an increase in the efficiency of EET. NAD(H/+) are also pivotal cofactors participating in most redox biochemical reactions in the microbial metabolism, in which NADH provides cells with reducing equivalents for reductive reactions, meanwhile, NAD^+ is the electron acceptor upon the NAD^+-dependent oxidative breakdown of organic substrates. In addition, NAD^+ is also linked intrinsically to signaling reactions inside and outside cells that control gene expression, Ca^{2+} mobilization, cell death and aging, etc. Cofactor engineering was thus developed to manipulate intracellular NAD (H/+) level via metabolic engineering approaches, which had been proven to enable an efficient control of intracellular NAD(H/+) availability. Such cofactor engineering strategies mainly included introducing biogenesis of NAD^+ to control total NAD (H/+) level and manipulating the consumption and regeneration of NADH to regulate the $NADH/NAD^+$ ratio. The underlying EET mechanisms of *Shewanella oneidensis* MR-1, one of the most intensively studied metal-reducing exoelectrogens, were intensively investigated in recent decade. Electrons were generated from central carbon metabolism of lactate (i.e., the carbon source and electron donor to *S. oneidensis*) and stored in the form of NADH as the intracellular electron carrier, which was converted from NAD^+ via oxidative reactions catalyzed by a number of dehydrogenases. Electrons subsequently transferred from NADH to the outer membranes through the menaquinol pool and a number of conductive c-type cytochromes (c-Cyts) including OmcA-MtrCAB and CymA and finally to the extracellular electron acceptors such as carbon electrodes or metal minerals through the contacted-based EET by c-Cyts and the solvable electron shuttle-mediated EET by redox active flavins. Thus, NADH and NAD^+ were not only essential cofactors for the metabolism of S. oneidensis, but also the intracellular electron carrier and source for its EET. However, there was barely any cofactor engineering effort in genetic manipulation of intracellular NADH to study its effects on EET of *S. oneidensis*.

In the electrochemically active *C. ljungdahlii*, nicotinamide adenine dinucleotide (NAD^+) and its reduced form NADH are essential cofactors and electron carriers that are primarily involved with cellular metabolic reactions and energy production. The ratio of $NAD^+/NADH$ represents the intracellular redox state in the cells. It has central functions in extracellular electron transfer and metabolic pathways. Cofactors are involved in most redox reactions and other metabolic reactions in electroactive organisms. NAD^+ and NADH are constantly regenerated with catabolism and anabolism. The ratio of $NADH/NAD^+$ has significant influence on energy transformation and the enzymatic activity of several. As electron carriers, $NADH/NAD^+$ also play important roles in MESs. In order to increase MFC performance, new and feasible cofactor genetic engineering approaches can be applied in MET to change intracellular redox state, which affects MFC electricity generation. For example, *P. aeruginosa* by $NAD^+/NADH$ cofactor manipulation shows a 50% increase in current density compared with the unmodified strain. Overexpression of the *Nad* gene, which adjusts the NAD^+ metabolic pathway was shown to increase the electricity production in MFC. *C.ljungdahlii* can generate a current in the MFC with the redox mediator methylene blue and formate dehydrogenase is the key enzyme that catalyzes the NADH biosynthesis pathway in

Chapter 3 Enhancing Extracellular Electron Flux

them. NAD⁺ dependent formate dehydrogenase (*Fdh*) from *Candida boidinii* could convert 1 M of formate and NAD⁺ into 1 M of NADH and CO_2. Genetic engineering of *E. coli* strains by the overexpression of an NAD⁺ dependent formate dehydrogenase that could increase NADH or NAD⁺ has been reported in the patents. Similarly, NAD⁺ dependent formate dehydrogenase, encoded by *fdh* gene, has always been a research hotspot in the area of NADH regeneration. From the above considerations, if NADH regeneration is applied in *C.ljungdahlii* by molecular biological technology, enough NADH can be accumulated that can be used for *C. ljungdahlii* MFC electricity.

Nicotinamide adenine dinucleotide (NAD⁺) and its reduced form NADH are essential cofactors for the metabolism of microorganisms, which are also hypothesized to be the intracellular electron pool for EET, enabling bioelectricity production of exoelectrogens. Previous efforts focused on increasing the intracellular [NADH]/[NAD⁺] ratio by means of metabolic engineering methods, thus enhancing the EET rate. To enhance EET in in *S. oneidensis*, metabolic engineering strategy were adapted to engineer and drive the metabolic flux towards the enhancement of intracellular NADH regeneration and the [NADH]/[NAD⁺] ratio, which subsequently increased the EET rate. In addition to the regulation of the intracellular [NADH]/[NAD⁺] ratio, increasing the total intracellular level of NAD(H/+) (i.e., total NADH and NAD⁺) via enhancing NAD⁺ de novo biosynthesis is another method to increase the EET rate that has been largely neglected in the past.

In other method increase of electron flux in EET can be increased by the oxidation of NADH there by making the NAD⁺ available for cellular metabolism. This strategy can also increase the rate of substrate oxidation in cell. This can be achieved by expressing the enzyme which oxidise NADH at high rate. NADH dehydrogenase II is one such enzyme which can be overexpressed to enhance EET (Fig 3.1).

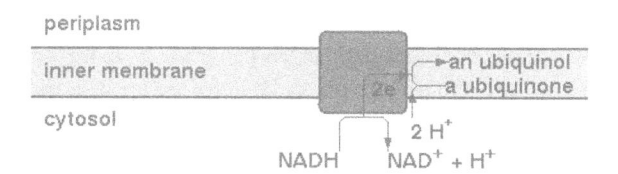

Fig 3.1: Mechanism of NADH dehydrogenase-II [It accepts electrons from NADH and donates them to quinone cycle without pumping protons from cytosol to periplasmic space].

Type-II NADH:quinone oxidoreductases (NDH-2s) are membrane proteins involved in respiratory chains and the only enzymes with NADH:quinone oxidoreductase activity expressed in *Bacillus subtilis*, one of the exoelectrogenic bacteria with unknown EET mechanism. NDH-2s are membrane proteins involved in respiratory chains and in the metabolic regeneration of NAD^+. These enzymes belong to the two-dinucleotide binding domains flavoprotein superfamily (tDBDF), which share two structurally related Rossmann fold domains for the binding of dinucleotides (FAD and NADH). Structural studies on NDH-2 have long contributed to the understanding of this enzyme, solving several controversies including the number and location of the substrate binding sites. Functional information on NDH-2 is still not fully understood and its role and the reaction mechanism of NDH-2 remain unknown (Fig 3.2). One significant property associated with NDH-2 is its ability to uncouple NADH oxidation and proton transport across the membrane which otherwise can lead to the development of high proton motive force that can, in turn, impede respiratory process as in case of NDH-1. Expression of NDH-2 can lead to higher metabolic rates and especially when there is high energy available to the cell (Gyan et al., 2006).

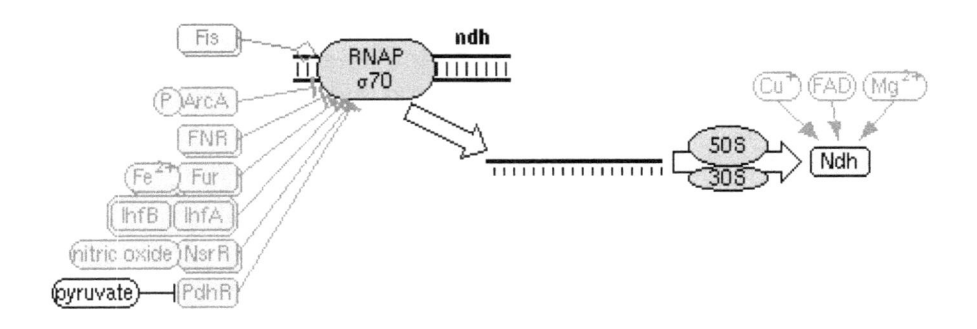

Fig 3.2: Regulation of *ndh* gene (*E. coli*) and its expression (Keseler et al., 2017).

It is well understood from the earlier studies that an appropriate ratio of electron carriers like NAD can play an important role in MET (Han et al., 2016). NAD pool (both oxidized and reduced forms) represents major the intracellular pool of electron for respiratory electron transport chain and EET(Berríos-Rivera et al., 2002b). In this context of increasing the performance of MES, we have focused on enhancing extracellular electron transport in microbial electrochemical systems by genetic engineering of NDH-2 into *E.coli* cells which can increase the rate of NADH oxidation there by increasing the electron flux into the electron transport chain and EET.

3.2 Experimental
3.2.1 Strains and growth conditions

All strains, unless otherwise specified, were grown overnight at 37 °C with 250 rpm shaking. Then overnight cultures were back-diluted into LB media at a dilution of 1:100 and grown with 250 rpm shaking till an OD of 0.8 is reached. Expression of the *ndh* gene was induced by addition of isopropyl β-D-1-thiogalactopyranoside (IPTG) at a concentration of 1 mM to the media. Cells were further grown for 3 hours before being harvesting. For bioelectrochemical system operation and iron reduction assays, induced cells were washed with M9 minimal salts solution and grown in M9 minimal media supplied with Wolfe's vitamins, minerals and appropriate antibiotics, as necessary.

3.2.2 Cloning and expression of NDH-2

The *ndh* gene (BSU12290 or NCBI gene ID: 939832) of *Bacillus subtilis* (Genbank accession number: KX470414.1) that has been isolated from the long term operated BES reactor was amplified using synthetically designed primers (Table 3.1) with BamH1 and HindIII enzyme restriction sites at the downstream of start and stop codons. The amplicon with the desired gene sequence and pET-28a(+) (Novagen) with a poly-histidine-tag at its N-terminal were subjected to restriction digestion with high fidelity BamH1 and HindIII (NEB, USA) restriction enzymes at 37°C for overnight followed by ligation with T4 DNA ligase (NEB, USA) at 16°C overnight. Further, recombinant vector with *ndh* gene is transformed into the *Escherichia coli* BL21 (DE5) cells for protein expression and stored as glycerol stocks at -80°C.

Table 3.1: Primers used for cloning and expression of ndh gene in this study.

Gene	Enzyme	Primer sequence (5` to 3`)
ndh	NADH dehydrogenase II	GCAGGAGGATCCATGTCAAAACATATTGTC
		ATATGAAAGCTTAGGTAAGCCAGGCTGAA
For cloning and expression of ndh gene		
Gene	Enzyme	Primer sequence (5` to 3`)
ndh	NADH dehydrogenase II	GCAGGAGGATCCATGTCAAAACATATTGTC
		ATATGAAAGCTTAGGTAAGCCAGGCTGAA

The transformed cells were then used for expression of *ndh* gene in two stages according to the protocol described above. The induced culture was further incubated to 3 hrs and then harvested in centrifuge at 4°C (8000 rpm/5min). Cell pellet was further processed based on downstream application.

3.2.2.1 General DNA techniques: All DNA manipulations were carried out by the standard procedures as described previously (Sambrook et al., 1989). All enzymatic manipulations involved reagents from NEB (Alkaline phosphatase, T4 DNA ligase, Polynucleotide kinase etc). For cloning, mutagenesis and sequencing PCR, desalted primers were synthesized from Integrated DNA Technologies (IDT) and PAGE-purified primers from Sigma were used for qPCR and qRT-PCR analysis.

3.2.2.2 Isolation of plasmid DNA: Plasmid isolation was done using Qiagen spin columns and procedure followed as they recommended. Briefly, 5ml of overnight grown culture was harvested by centrifugation; cell pellet was lysed and treated with alkaline lysis buffer. The solution is neutralized by neutralization buffer and adjusted to high salt binding conditions. The mixture was centrifuged and supernatant was applied to mini spin column. The resin was washed with wash buffer and plasmid DNA was eluted in an elution buffer (10 mM Tris-0.5 mM EDTA, pH 8.0).

3.2.2.3 Restriction digestion of DNA samples: Restriction digestion was carried out with NEB (New England Buffer) BamHI–HF and HinDIII-HF restriction enzymes and reaction protocols were followed according to the manufacturer's recommendations. Mostly, reactions were carried out at 37^0C for specified time and the samples were heat treated to inactivate the enzyme. Finally the samples were analysed on agarose gel electrophoresis to verify the digestion of DNA samples.

3.2.2.4 Agarose gel electrophoresis: DNA fragments and linearised plasmid DNA was resolved on 1-1.7% agarose gels for gel extraction and analytical purposes. Gel loading buffer was added to the samples, loaded onto the gel and electrophoresis was carried out in 1X TAE at 10 v/cm of gel. Ethidium bromide (0.5 μg/ml was used to stain the DNA fragments and visualized on a UV transilluminator (Syngene). The size of the DNA samples was calculated in reference to the standard markers (Thermo Scientific).

Table 3.2: Buffers and solutions used for agarose gel electrophoresis of DNA samples.

Buffers/Solutions	Composition
TE buffer	10 mM Tris-Cl, pH 8.0 and 1.0 mM EDTA in deionized water
50X TAE	242 g Tris base; 57.1 ml glacial acetic acid; 100 ml 0.5 M EDTA, pH 8.0 were mixed with and final volume was made upto 1L with deionized water
6X DNA loading buffer	0.25 % bromophenol blue; 30 % glycerol dissolved in deionized water

3.2.2.5 Gel purification of DNA fragments: After electrophoresis, DNA fragments were excised from the gel and transferred to microcentrifuge tube. DNA was extracted using Qiagen gel extraction columns following the manufacturer's protocols. DNA fragments were solubilized in 3 volumes of QG buffer and one volume of isopropanol was added. The mixture was applied to gel extraction columns, washed and finally DNA was eluted in 1XTE buffer or water.

3.2.2.6 Polymerase chain reaction: Polymerase chain reaction (PCR) was performed for amplification of DNA fragments in a 50 µl of reaction volume containing 1X PCR buffer, 250 mM of deoxy-ribonucleoside triphosphates (dNTPs), 20 pmol of upstream and downstream primers and 1U of Q5 high fidelity DNA polymerase enzyme (NEB). Generally, amplification was performed for 35 cycles (denaturation at 95^0C for 2 min, annealing at 60^0C for 30 seconds and elongation at 72^0C for 1min/kb of template) followed by final elongation for 8 min. The reactions were performed in a PCR thermocycler (Eppendorf).

3.2.2.7 Ligation and double digestion: Restriction digested insert and plasmid were ligated with T4 DNA ligase (NEB) at ratio of 1:3 vector to insertaccording to the manufacturer's recommendations. Reaction was carried out at 4°C for overnight and the ligation mixture was directly used for cloning. To confirm the insertion of *ndh*into the peT28a, ligation mixture was double digested using BamHI and HinDIII for overnight and the release of insert was checked using 1% agarose gel electrophoresis.

3.2.2.8 DNA sequencing: For sequencing of DNA fragments and clones, PCR amplification was done using specific primers for 24 cycles (denaturation at 96^0C for 30 seconds, annealing at 50^0C for 15 seconds and elongation at 60^0C for 4 min per cycle) by using reagents from Applied Biosystems. Sequencing was carried out on automated sequencer (Applied Biosystems) at Bioserve facility, Hyderabad. The readouts were analyzed using chromas software.

3.2.2.9 Preparation of competent cells and transformation: Protocol for preparing E .coli competent cells using $CaCl_2$ was described previously (Sambrook et al., 2001). Briefly, 100 ml of *E.coli* culture was grown till OD600 0.3-0.4 and the flask containing culture was incubated on ice for 30 min. Cells were harvested by centrifugation at 4000 rpm at 4^0C for 10 min. The pellet was resuspended in chilled 50 mM $CaCl_2$ and incubated on ice for 30 min. Then, the cells were harvested and resuspended in 50 mM $CaCl_2$ with 20% glycerol. Aliquots were prepared and stored at -80^0C. For transformation, 50 ng of super coiled plasmid DNA or ligation mixture was added to thawed competent cells, incubated for 30-40 min and heat shock was given at 37^0C for 2 min. Next, 1 ml of LB media was added to the cells and incubated for growth at 37^0C. Transformed colonies were selected by plating the cells on agar plates containing appropriate antibiotic.

3.2.3 Colony PCR

Clones observed after overnight growth were randomly picked and subjected to lysis by heating them at 95°C in nuclease free water. Using lysate as template, PCR was performed using protocol mentioned in the PCR section before.

3.2.4 Protein purification

The cells lysis of 2 litre culture, grown and harvested as above was carried out by resuspending the cell pellet in 100ml of lysis buffer (50 mM NaCl, 2 mM $MgCl_2$, 0.25 mg/ml lysozyme, 2 mM pefabloc, 2µl benzonase) followed by incubation at room temperature for 30 min. Further, the cells were sonicated with a pulse rate of 2 seconds on and 3 seconds off cycle (50% amplitude) for 5 min on ice, after which the lysate is centrifuged at low temperature (10,000 rpm/30 min) to remove cell debris. The supernatant was then transferred slowly to another tube leaving the cell pellet undisturbed that retain cell debris. To separate the membrane proteins from cytosolic proteins, the supernatant collected in the above step is subjected to ultracentrifugation at 40,000 rpm/50 min (4^0C) and the resultant pellet holding membrane proteins was washed in 25 ml of solubilization buffer (50 mM Na_2PO_4, 500 mM NaCl, 5% glycerol, 2mM pefabloc) and then ultracentifuged in the fresh tube following the similar conditions. The pellet is then resuspended in 25 ml of solubilization buffer with 6 ml of n-dodecyl- β-D-maltoside (DDM) detergent (10 % stock) and then ultracentrifuged at 40,000 rpm/50 min after its incubation for 2 hrs at 4°C on a test tube rocker. Immediately after the spin, the supernatant containing membrane fraction was transferred into a fresh tube without disturbing the pellet to avoid mixing of insoluble proteins. His-tagged membrane bound NDH-2 was further purified through affinity chromatography by making use of Talon beads (Clonetech, Japan). The membrane suspension was incubated with the equilibrated resin at 4°C for 1 hr to enhance the binding of NDH-2 and is then transferred to a gravity flow column. The unbound proteins in the column is then washed with 10 column volumes of wash buffer (50 mM Na$_2$PO$_4$, 500 mM NaCl, 5% glycerol, 0.2% DDM), after which the NDH-2 bound with resin was eluted with a gradient of 5-100% of immidazole (500 mM imidazole = 100%) in the buffer and collected in multiple fractions. The column purified protein is then dialyzed to remove imidazole. The fractions collected during the chromatography were loaded on to the SDS-page gel (12% - separating gel) to check for the purity (Table 3.2).

Table 3.3: Buffers and solutions used for SDS-PAGE analysis of proteins.

Buffers/Solutions	Composition
30 % Acrylamide	29 g Acrylamide and 1.0g bis-acrylamide was dissolved in 100 ml of deionized water
Tris (pH 8.8)	18.17 g Tris base was dissolved in 100 ml of deionized water after adjusting pH to 8.8 using 6N HCL
Tris (pH 6.8)	6.06 g Tris base was dissolved in 100 ml of deionized water after adjusting pH to 6.8 using 6N HCL
5 X SDS sample buffer	1.5 M Tris-Cl pH 6.8; 5.0 % SDS solution; 25 % glycerol; 12.5 % β-mercapto ethanol, 0.06 % bromophenol blue was dissolved in 10 ml of deionized water
Laemmli buffer	3 g Tris; 14.4 g Glycine and 1.0 g SDS was dissolved in deionized water and volume was made up to 1L
Gel staining solution	50 % methanol; 10 % acetic acid and 0.25 % Coomassie brilliant blue (R250) in deionized water
Gel destaining solution	40 % methanol; 10 % acetic acid in deionized water

Further, the protein confirmation was done through western blotting by transfer of proteins separated during the electrophoresis onto the PVDF membrane through electro-blotting (25 V) for overnight at low temperature. Later, the membrane is blocked with 2% of blocking agent and then incubated at 4°C with mouse anti-his primary antibodies, followed by incubation with HRP linked secondary antibody and visualized through luminal based detection.

Table 3.4: Buffers and solutions used for western blotting.

Buffers/Solutions	Composition
10 X PBS	80 g NaCl, 2 g KCl, 15 g Na$_2$HPO$_4$, 2.1 g KH$_2$PO$_4$ were dissolved in 1L of deionized water and pH was adjusted to 7.4 with 1N HCl
Transfer buffer	30 mM Tris-Cl (pH 7.5), 240 mM glycine and 10 % methanol was dissolved in deionized water
Blocking buffer	1 X PBS, 0.5 % skimmed milk. 0.1 % Tween-20
Wash buffer	1 X PBS, 0.5 % Tween-20

3.2.5 Bioelectrochemical system design and characterization

The harvested cells were washed thrice with M9 minimal media before its use in BES and resuspended in the same, supplemented with wolfe's trace metals, vitamins and 20 mM lactate as a carbon source. Cells were adjusted to OD of 1 and purged with O$_2$ free N$_2$:CO$_2$ (80:20) mixture prior to its inoculation into BES. The role of NDH-2 in EET was evaluated using single chambered BES with a working volume of 60 ml using FTO plate anode with a surface area of 6 cm^2, platinum wire as cathode and Ag/AgCl (3.5 M KCl) as a reference electrode (Fig 3.3). Culture preparation and reactor setup was carried in the anaerobic glove box (Coy Laboratories) supplied with oxygen free mixture gas containing nitrogen (85%), carbon di oxide (10%) and hydrogen (5%).

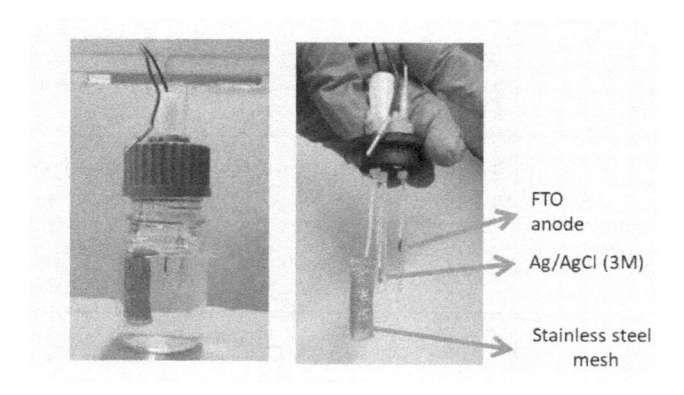

Fig 3.3: Electrodes and reactor setup used for the bioelectrochemical studies.

3.2.6 Bioelectrochemical analysis

To the anode of MES, +200 mV of potential was applied using multi-channel potentiostat-galvanostat system (Biologic VMP-3, France) to enhance the bacterial growth on the anode (biofilm formation) and then the current generated was measured (Fig 3.4).

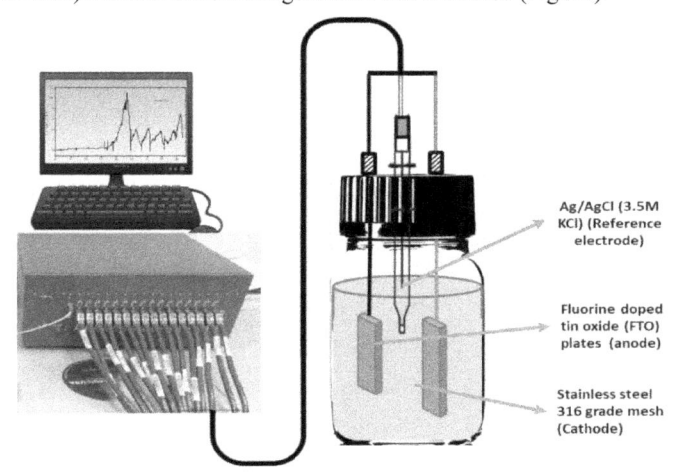

Fig 3.4: Three electrode electrochemical setup used for the study.

Cyclic Voltammograms were recorded for both induced and uninduced cultures at 1 mV/s scan rate in a scan range of 0.5 to -0.5 V using VMP-3potentistat. To understand the growth of biofilm on the electrode, Potentiostatic Electrochemical Impedance Spectroscopy (PEIS) was performed in a frequency range of 100 MHz to 10 mHz and a potential of + 200 mV was applied on to the anode during the recording of the technique. All the potentials discussed in the current work are with respect to the Ag/AgCl, KCl (3.5 M) reference electrode unless stated otherwise.

3.2.7 Estimation of biofilm on the anode by biofilm viability kit

The biofilm formation on the electrode after the BES operation was evaluated for respective reactors using Film tracer LIVE/DEAD Biofilm Viability Kit (Invitrogen, USA). The anodes from both the reactors were removed as soon as the reactor was opened and washed with phosphate buffer to remove the unbound and loosely bound cells. The biofilm was then stained with SYTO-9 and propidium iodide dyes mixture as suggested by the manufacturer, followed by incubation at room temperature for 30 minutes in dark. The electrodes were then observed under the epifluorescence microscope (Nikon eclipse *80i*, Japan) using DAPi and CY3 filters after washing them thoroughly with filter sterilized water.

3.2.8 NAD/NADH estimation

The ratio of NAD/NADH in both induced and uninduced cultures was estimated using NAD$^+$/NADH Quantitation Kit (Sigma Aldrich). The harvested cells from both induced and uninduced cultures were washed thoroughly with chilled phosphate buffer and extraction of NADH/NAD$^+$ was performed in NADH extraction buffer by 2 freeze thaw cycles in liquid nitrogen for 20 minutes (10 min in Liquid N$_2$ and 10 minutes at room temperature). Debris was removed by centrifugation at 13,000 g for 10 min and the resulting supernatant containing NAD/NADH was deprotinized by passing it through 10 kDa protein cutoff spin filter to remove NADH consuming proteins. Deprotinized supernatant was used in a NAD$^+$ cycling assay to determine the amounts of NADH and total NAD content following the manufacturer's instructions. The colorimetric estimation of the NADH in the sample along with standards was done using Multimode plate reader (SpectraMax M2e, Molecular Devices) at a wavelength of 450 nm. All the steps were carried out at low temperature in order to maintain stability of NADH.

3.2.9 Iron utilization assays using Ferrozine assay

The role of NDH-2 in extracellular electron transport was evaluated through iron reduction assay (ferrozine assay). Initially the cells were grown in LB broth, following the induction of NDH-2.Cells from both induced and uninduced cultures were pelleted and washed with M9 minimal media. Washed pellet was resuspended in M9 minimal media separately with lactate as electron donor and ferric citrate (10 mM) as electron acceptor at an OD of one in anaerobic glove box (Coy, USA). The kinetics of iron reduction [Fe(II)] in both the cultures was studied by Ferrozine assay (Ruebush et al., 2006).

3.2.10 Biohydrogen production

To understand effect of NDH-2 on biohydrogen production, cells were grown as mentioned in the section 3.2.9 and finally inoculated in M9 minimal media supplied with vitamins and minerals. Cells were allowed to grow for 48 hours at 37°C and the gas produced was analyzed at 48th hour using gas chromatography with argon as carrier gas, the injector and detector at 60°C and the oven at 40°C (GC; Agilent technologies) using thermal conductivity detector (TCD) with Heysep Q column.

3.2.11 Spectroelectrochemical studies

Spectroelectrochemical experiments were carried out at room temperature in an anaerobic glove box to prevent the leakage of oxygen into the cell. All experiments were performed using three electrodes setup with platinum mesh as working electrode (WE), a platinum wire as counter electrode (CE) and a Ag/AgCl, KCl (3.5M0 reference electrode (RE). Biologic potentiostat/galvanostat VSP electrochemical work station (Biologic science instruments, France) coupled with a using SEC2000-UV/VIS spectrometer (ALS Co., Ltd, Japan).Removal of oxygen from the cell was achieved by using nitrogen gas flow into the cell. After a few minutes, approximately 350 µL of the protein sample, which was previously made anaerobic in a separate vial, was then injected into the electrochemical cell. The protein concentrations were in the 1–5 µM range and the redox mediator concentrations were between 1 and 100 µM.

3.3 Results and Discussion
3.3.1 Cloning and Protein purification

NDH-2 is non-proton pumping type II NADH dehydrogenase which plays a central role in the respiratory metabolism of bacteria. Its ability to overcome the inhibition posed by high proton motive force makes it an ideal protein to study to increase the metabolic and electron flux through the cellular metabolism and electron transport chain.

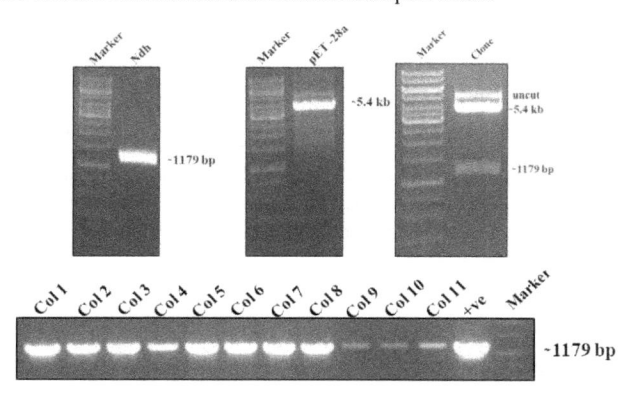

Fig 3.5: Double digestion of insert and vector for confirmation of insertion of gene into vector and Clones conformation using colony PCR for positive colonies

To understand the effect of NDH-2 on extracellular electron transport, the gene encoding NDH-2 was transformed into *E. coli* BL21 (DE3) using pET28a vector under the T7 promoter. Primers were designed in both 5' and 3' ends by inserting BamHI and HinDIII restriction enzymes sites. The '*ndh*' gene was amplified from *Bacillus subtilis* (isolated from lab MFC rectors) whole genome. Restriction digestion of both insert and vector were carried out and ligated using T4 DNA ligase. The ligated product was added to *E. coli* BL21 ultra competent cells and transformed by following procedure described in the materials. The insertion of *ndh* gene into the vector was confirmed by double digestion using BamHI and HinDIII restriction enzymes. Gene orientation of *ndh* was confirmed by the sequencing of the recombinant vector. Clone confirmation of transformed cells was done by using colony PCR (Fig 3.5). Optimum expression of NDH-2 by engineered cells was standardized at different temperatures (including 37 °C, 28 °C and 18 °C) and IPTG (isopropyl β-D-1-thiogalactopyranoside) (0.3, 0.6 and 1.0 mM) conditions to obtain protein expression in the soluble fraction. Optimum protein expression in soluble fraction was observed when induced at 37°C with 1 mM IPTG for three hours. Cells expressing NDH-2 protein were light greenish in color compared to the cream color of uninduced pellet (Fig. 3.6) indicating the expression of the NDH-2 protein. Green color of the protein is due to the presence of non-covalently bound FAD cofactor in the NDH-2 protein (Heikal et al., 2014). The pI and Mr of the proteins were predicted using ExPASy tools accessed from cn.expasy.org/. Table 3.5 summarizes the predicted properties of NDH-2 based on the protein sequence.

Table 3.5: Properties of NDH-2 protein based on protein sequence information (using ExPASy).

Molecular weight (Da)	41953.16
Number of amino acids	392
Isoelectric pH (pI)	6.28
Extinction coefficient (M^{-1} cm^{-1} in water)	30370
Abs 0.1% (=1 g/l)	0.724

NDH-2 induced cell lysate showed a distinct protein band close to 42 KDa which is absent in the uninduced fraction indicating successful expression of NDH-2 in transformed cells.

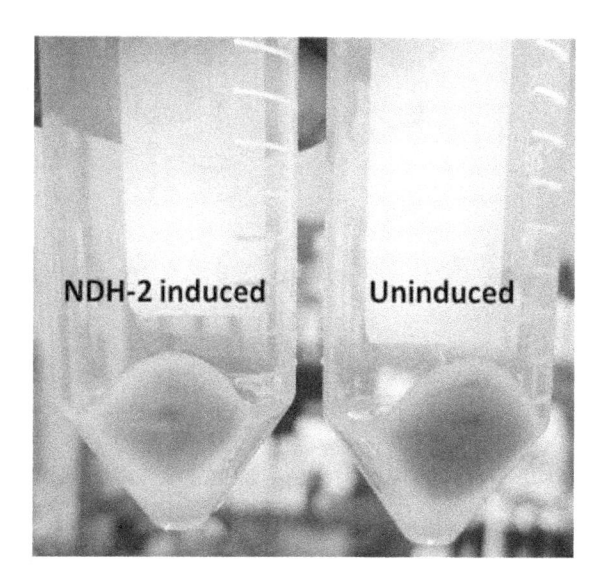

Chapter 3   Enhancing Extracellular Electron Flux

Fig 3.6: Cell pellets of NDH-2 induced (1 mM IPTG) and uninduced cultures [Green color of the pellet is due to the presence of FAD cofactor in the NDH-2 enzyme].

After its successful expression in *E. coli* BL21 cells, the His-tagged NDH-2 was then purified from total protein fraction by Ni-NTA affinity chromatography using his-tag of NDH-2 (Fig 3.7). Purity of the protein was checked on 12% SDS PAGE which showed good quantity of purified protein near 42 kDa (Fig. 3.8).

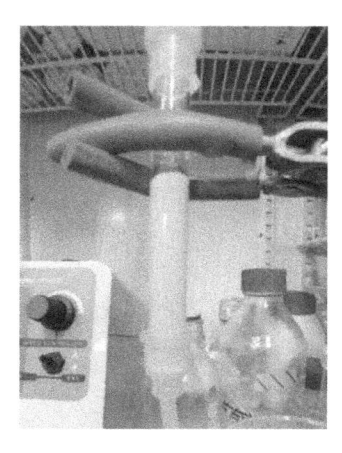

Fig 3.7: Bound NDH-2 to the Ni-NTA affinity purification column after binding step [Color of the column is due to the FAD cofactor].

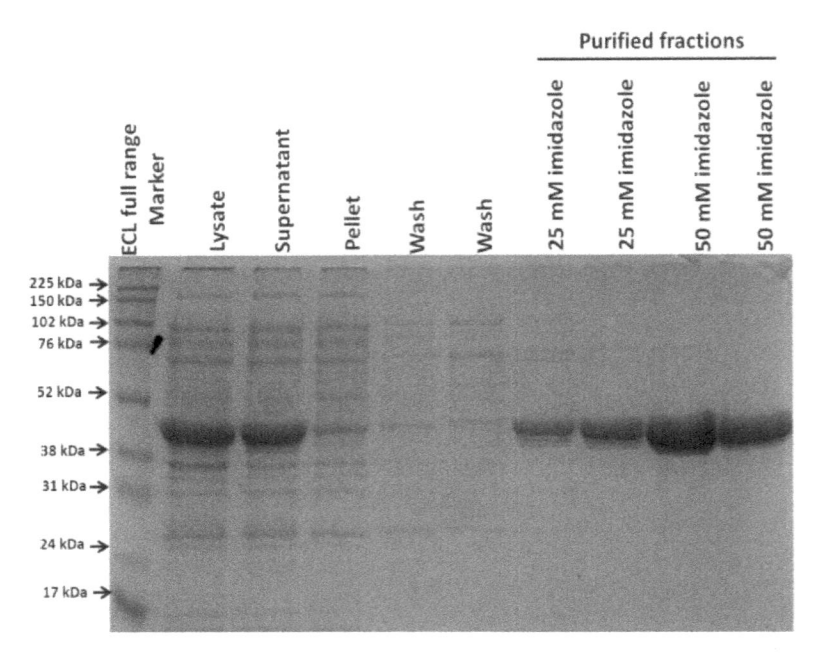

Fig 3.8: Protein purification using his tag purification (Ni-NTA beads). Purified protein was light greenish in color emphasizing the presence of FAD$^+$ cofactor in the NDH-2.

As the localization of heterologously expressed NDH-2 is important, western blotting with an anti his antibody has been performed which has produced a band at the expected molecular mass of NDH-2, i.e., nearly 42 kDa, indicating proper expression and localization of NDH-2 (Fig 3.9).

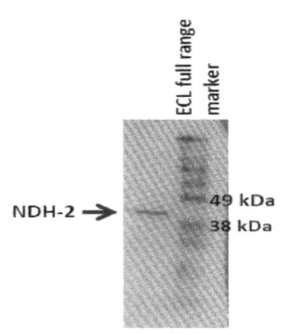

Fig 3.9: Western blotting of the NDH-2 protein using anti-6Xhis antibodies. Presence of band at ~42 KDa indicates the expression of the NDH-2 in the membrane fraction.

Redox activity of the NDH-2 was probed by using UV- Vis spectroscopy. Visible spectra of oxidized NDH-2 have showed a sharp peak at 450 nm indicating the presence of FAD^+ cofactor in the enzyme(Salewski et al., 2016).

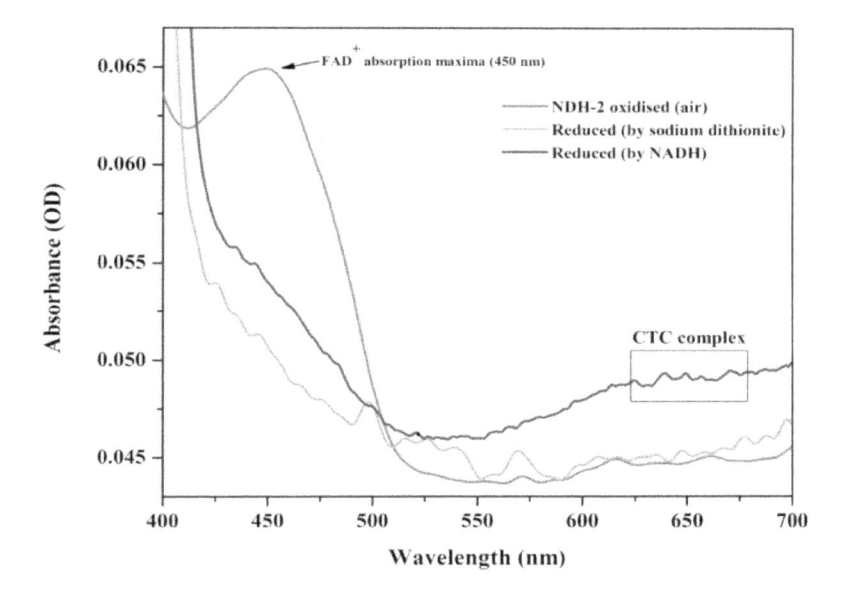

Fig 3.10: Absorption (visible) spectra of purified NDH-2 under oxidizing (air) and under reducing conditions (reduced by sodium dithionite and NADH). Addition of dithionite has resulted in reduction of FAD^+ which lead to the over decrease in absorbance at 450 nm. Charge transfer complex (CTC) formation can only be seen in case of NADH.

Reduction of NDH-2 using sodium dithionite and NADH has resulted in gradual decrease in 450 nm peak which finally diminished indicating the reduction of FAD^+ to $FADH_2$ by sodium dithionite and NADH which has no absorbance at 450 nm (Fig 3.10). In case of NADH reduction, a charge transfer complex or CTC (intermediate enzyme, NAD complex) was observed at ~650 nm (Fig 3.10). While in case of sodium dithionite, no CTC formation was observed indicating that it has the ability to reduce FAD^+ but lacks the ability to form CTC which is essential for the enzyme activity (Blaza et al., 2017).

3.3.2 Bioelectrogenesis of NDH-2 clone

The NDH-2 clone cells were initially grown in LB media overnight from glycerol stocks. Overnight grown cells were diluted into the fresh LB broth at 1:100 ratios. Once the cells are induced at 0.8 OD and further grown for 3 hours aerobically, cells were pelleted and washed thrice with nitrogen purged M9 minimal media to remove any carryover media components and dissolved in fresh Minimal media supplied with sodium lactate, vitamins and minerals. Cells were adjusted to an OD of 1.0 while dissolving the pellet into the media. After transferring the NDH-2 cells into reactor, media was purged with oxygen free nitrogen, carbon dioxide mixture (80% : 20%, V:V) for half an hour to remove any traces of gases. Chronoamperometry was not performed during the first hour, as cells would take time to synthesize the machinery required for anaerobic growth and lactate utilization. Current producing ability of the NDH-2 induced cells was compared with uninduced cells where NDH-2 overexpression was absent. An applied potential of + 0.2 V was continuously applied on to the anode of both the systems to promote the growth of bacteria on anode (biofilm formation)(Liu et al., 2008).

3.3.3 Chronoamperometry (CA)

CA was performed for three cycles (each cycle comprising twenty four). In both the cases current production was stabilized after ~5 hours of starting the cycle (Fig. 3.11).

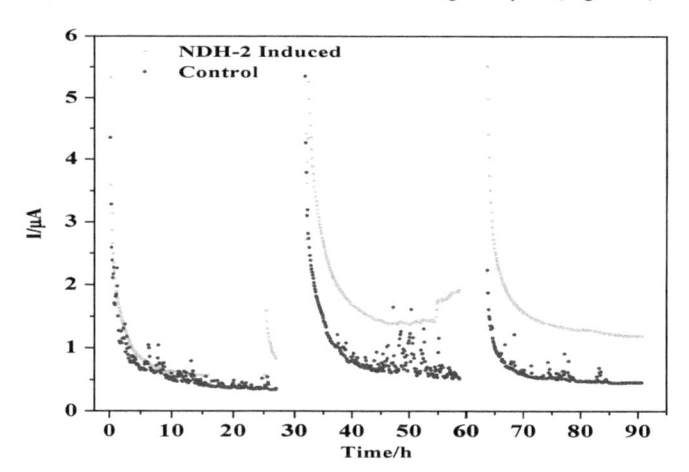

Fig 3.11: Bioelectrogenic activity of NDH-2 induced and uninduced BES reactors poised at + 0.2 V. NDH-2 induced culture has shown 3 fold increases in current production compared to uninduced MES reactor.

During the initial cycle, the NDH-2 BES reactor (0.6 µA) has only showed marginal increment in current production compared to the uninduced BES reactor (0.4 µA). While in the second and third cycle, NDH-2 BES (~1.9 µA & 1.6 µA) has showed significant increase in current production compared to uninduced BES (~0.6 µA & 0.6 µA) which is almost 3.16 folds increase in current production.

3.3.4 Cyclic voltammetry (CV)

CV was performed using three electrode configurations at low scan rates (1 mV.S^{-1}). To rule out the noninvolvement of exogenous mediators or media components, voltammograms with uninoculated anode and fresh media were generated. It was observed that, the uninoculated MES produced no catalytic current ruling out involvement of media components and exogenous mediators in current generation. CV's reveal microbial catalyzed redox reactions during substrate oxidation at the anode of MES and also redox reaction kinetics at bacteria-electrode interface. Cyclic voltammograms of NDH-2 induced MES and uninduced MES at various time points during operation along with comparison was presented in the Figure 3.12. The cyclic voltammograms of NDH-2 induced MES showed both oxidation (E_1) and reduction reaction (E_2) with maximum peak current in the reverse scan.

However, no considerable peak was observed with uninduced neither in forward nor in the reverse scan. Reduction peak (E_2) (with a midpoint potential of ~ -10 mV) at 6th hour of operation in case of NDH-2 BES was found to be lactate utilization (oxidation) peak which slowly shifted (peak splitting) towards reduction side (~ -70 mV by the end of 48th hour) of the voltammograms (Fu et al., 2013). Though the peak shifts were observed in previous *Geobacter* biofilm studies, they were mostly due to the diffusion limited conditions prevailing at thin biofilms due to the increasing scan rate where the diffusion of redox molecule to the electrode is not as fast as the electrode reduction which results in the peak shift (Richter et al., 2009). But in this case, the peak shift of E_2 was observed with respect to time at same scan rate of 1 mV. S^{-1} indicating that the peak shift generated was not as a result of the diffusion limited current. The plausible explanation for shift in formal potential of E_2 peak is the biofilm formation on electrode resulting in various micro-environments which can influence the formal potential of the redox molecule as the same redox mediator of electron transport chain will be exposed to different environments resulting in peak shift (Xiaoxi Chen et al., 2002). Another explanation for the peak shift seen in NDH-2 CV's is high concentration of redox molecules at thin biofilm resulting in lateral electron transfer to the adjacent redox mediators parallel to the surface of the electrode (Strycharz et al., 2011).

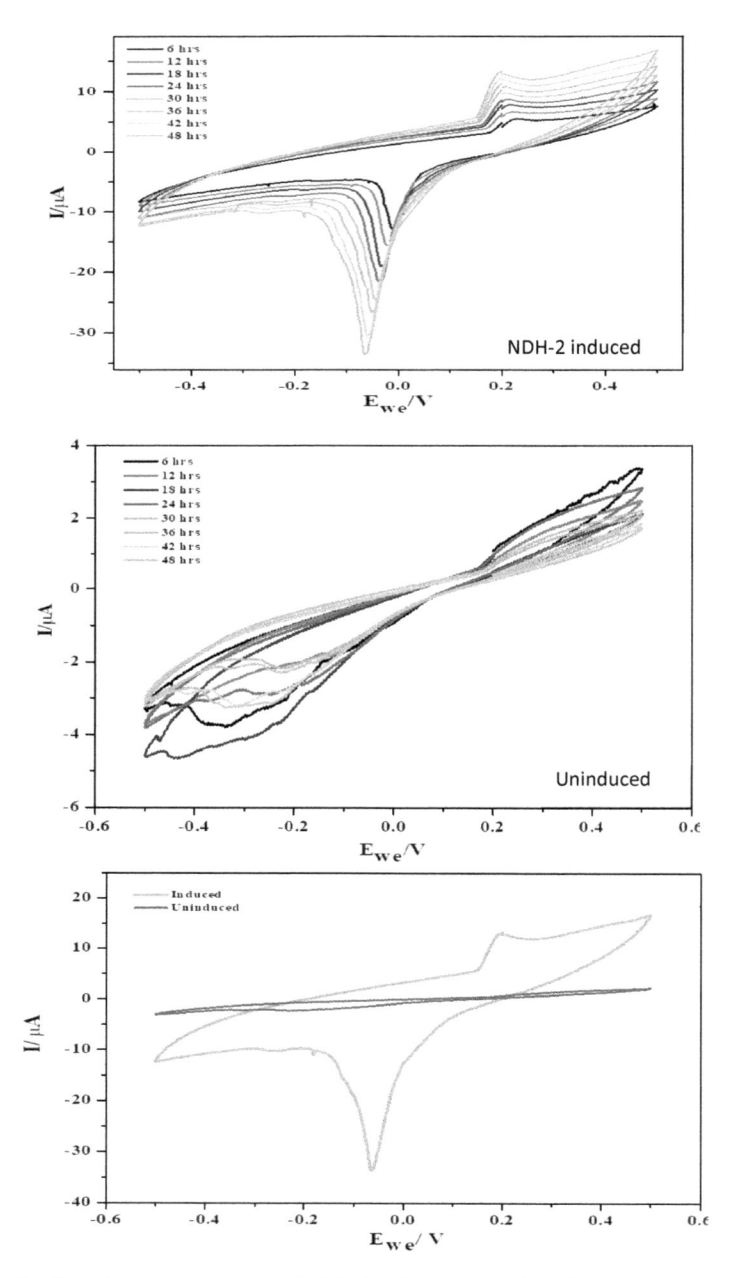

Fig 3.12: Cyclic voltammograms recorded during 48 hours period of time (at every six hours) at a scan rate of 0.001 V. s^{-1} (a) NDH-2 induced (b) Uninduced (c) Combined.

In this case, it can be understood that, NDH-2 expression has resulted in increased biofilm formation (discussed in section 3.3.5) which might have created microenvironment around the anode resulting in shifting of formal potentials of the redox compound.

3.3.5 Potentiostatic electrochemical impedance spectroscopy (PEIS)

PEIS was used to understand the biofilm formation and resulting changes in the charge transfer resistance (R_{CT}) to transport electrons from biofilm to the anode (Stöckl et al., 2016). Impedance measurements were performed at 100 MHz to 10 mHz frequency and a DC potential of + 0.2 V was applied during the measurement to enhance biofilm formation. Impedance of BES reactor with only media has shown huge resistance of ~ 500 kΩ indicating that media components have no role in decreasing the overall resistance of the cell. Results have demonstrated that the charge transfer resistance has decreased over time indicating the deposition of bacteria (biofilm formation) (Ben-Yoav et al., 2011; Ramasamy et al., 2008). NDH-2 induced BES has showed six fold decrement during 48 hours of operation (Fig 3.13). R_{CT} values decreased over time from ~ 40 kΩ at the beginning of the experiment to the ~14 kΩ by the end of the cycle. EIS plots of NDH-2 induced at various time points were presented in Nyquist diagrams (Fig 3.13). The decrease in impedance with time can be correlated to the formation of electroactive biofilm on anode which can reduce the activation losses (transfer of electrons from bacteria to the electrode) (Ben-Yoav et al., 2011). With the formation of biofilm on anode, the resistance for electron transport will decrease as direct electron transport will dominate the mode of extracellular electron transport in MES. While in case of uninduced MES, a different pattern of impedance plots were observed with respect to time of operation. R_{CT} values of parent strain nyquist plot suggest that the charge transfer resistance has increase initially and stabilized over time. At 12th hour of operation, R_{CT} has increased from ~138 kΩ to ~148 kΩ, where it got stabilized till 36th hour of operation. Though there was increase in R_{CT} at 42th hour (~ 162 kΩ), it finally returned to the previous stabilized state of ~145 kΩ suggesting a lower rate of biofilm formation in uninduced BES reactor compared to the NDH-2 induced.

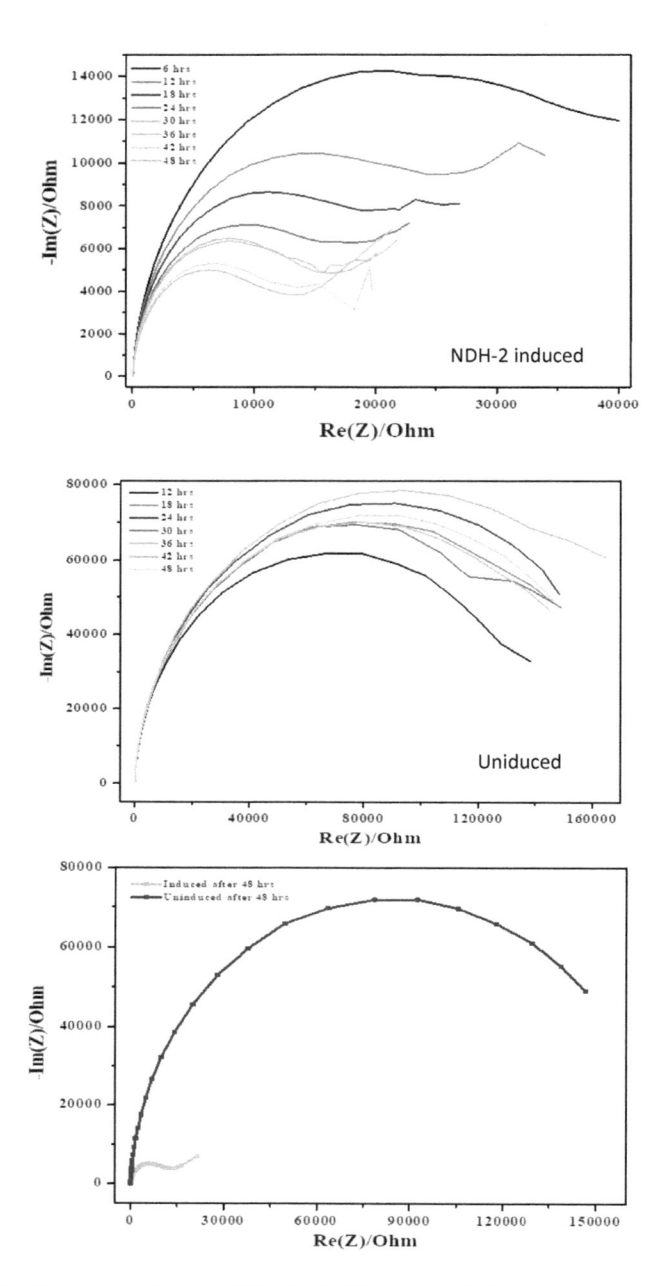

Fig 3.13: Biofilm monitoring by EIS (a) NDH-2 induced (b) Uninduced (c) Combined. Higher charge transfer decrement was observed in case of NDH-2 induced cells.

In addition to the EIS measurements, enhanced formation of biofilm in case of NDH-2 induced MES (compared to uninduced) was also confirmed by biofilm images taken at the end of the cycle using SYTO 9 fluorescent dye staining (Fig. 3.14). It can also be observed that the most of the cells are alive and healthy. Increased biofilm formation in NDH-2 cultures can be because of the excessive electron and metabolic flux generated by the NDH-2 cells due to the activity of NADH dehydrogenase 2. The excessive electrons generated might have favored the NDH-2 cells growth on anode to enhance direct electron transport enabling faster discharge of electrons from bacteria to anode.

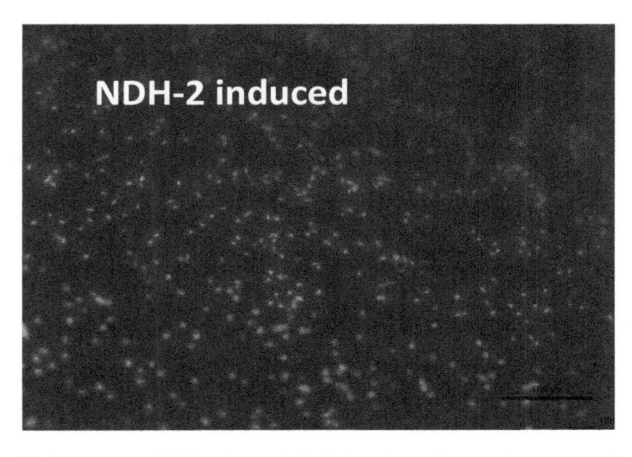

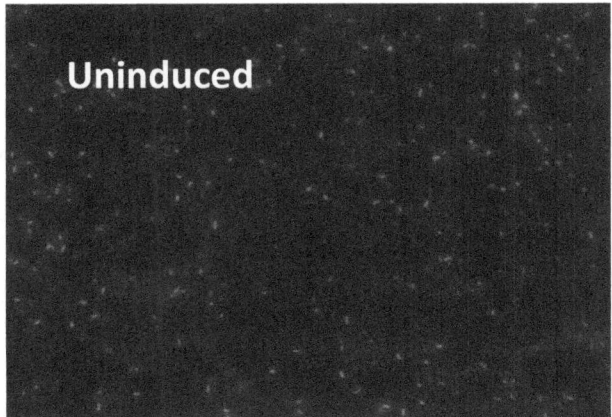

Fig 3.14: Biofilm formation on the FTO plates after impedance monitoring. Biofilm was stained using SYTO9 and propidium iodide dyes. After 48 hour cycle, it was observed that the cell density was more on the NDH-2 induced MES.

3.3.6 Effect of NDH-2 on NAD/NADH concentration, bioelectricity and biofilm development

Expression of *ndh* gene in *E coli* has resulted in 1.32 fold increase (~58.68 µM/ 10^9 cells) in total NAD (H) pool compared to the uninduced culture (~44.43 µM/10^9 cells) (Fig. 3.15). In NDH-2 induced cultures NAD^+ concentration was much higher than the NADH (< 2 µM/10^9 cells) indicating high catabolic rate as a result of NDH-2 activity in the cellular membrane (Berríos-Rivera et al., 2002a). Though the concentration of NADH in uninduced cells was negligible (< 2 µM / 10^9 cells) it was higher than the induced cells signifying the activity of NDH-2. It can be clearly seen that the bacteria has increased the production of total NAD/(H^+) to compensate for increased oxidation of NADH by NDH-2 enzyme.

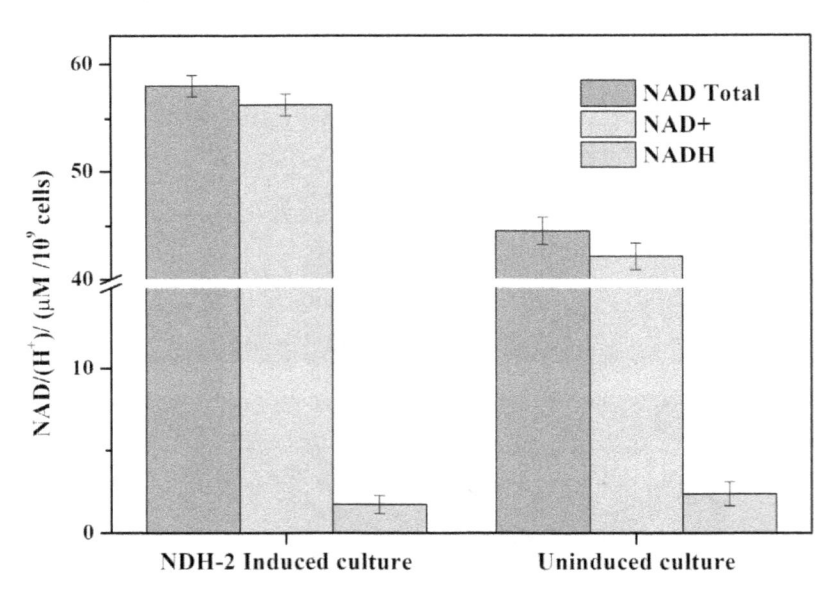

Fig 3.15: Total NAD^+, NAD^+, NADH concentration of aerobically grown cultures after 3 hours of NDH-2 induction. NADH/H^+ concentrations were found to be very low corresponding to the high rate of catabolism and increased electron flux into electron transport chain.

In addition to this, it has also been discovered recently that the enzyme NADH oxidase which catalyzes NAD^+ regeneration from NADH for maintaining glycolysis has a role to play in biofilm formation in *Streptococcus sanguinis* which suggests that the NAD^+ regeneration is one of the key factors in biofilm formation (Ge et al., 2016). NDH-2 being an enzyme which oxidizes NADH to NAD^+ might also play a similar role in biofilm formation as both the enzymes does equivalent function but in different perspectives. From the results it can be understood that intracellular NAD^+ concentration has substantially increased in NDH-2induced cells which might have triggered higher biofilm formation (Fig 3.14). The increased NADH oxidation has increased electrons released to EET pathway and played an important role in bioelectricity generation and overall electron flux through electron transport chain. In addition to this, positive effect of biofilm formation on the anode can be correlated with electrochemical impedance spectroscopy (EIS) results where it can be seen that the charge transfer resistance (R_{CT}) has decreased over time with increase in biofilm formation.

3.3.7 Biohydrogen production

To understand the role of NADH dehydrogenase 2 on biohydrogen production, induced and uninduced cell were grown separately in minimal media supplied with lactate as electron donor for 48 hours and gas composition was evaluated using gas chromatography (thermal conductivity detector (TCD)). NDH-2 induced cells has shown higher percentage of hydrogen gas (11.2 %) which is almost 2 fold higher than the uninduced cells which has only 5.7% of biohydrogen in total gas (Fig 3. 16). Increase in hydrogen production in NDH-2 cells can be attributed to the shift in redox balance because of activity of NDH-2. Increased oxidation of NADH due to activity of NDH-2 have resulted in faster regeneration of NAD^+ compared to the uninduced strain which has led to increased substrate conversion.

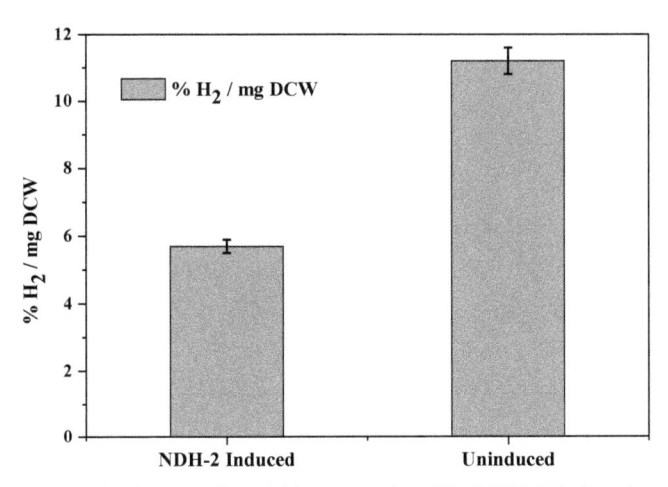

Fig 3.16: Percentage hydrogen of total biogas produced in NDH-2 induced and uninduced cells after 48 hours of growth.

3.3.8 Iron reduction by NDH-2 expressing cells

The capacity of NDH-2 induced cells to reduce the soluble iron forms was determined by using ferric citrate as the only final electron acceptor. In order to prevent the aerobic reduction of ferric citrate or oxygen acting as final electron acceptor to *E. coli*, media was flushed with oxygen free N_2-CO_2 mixture. It was found that the expression of *ndh* gene in *E. coli* has enhanced its ability to reduce the Iron (III) citrate with lactate as carbon under anaerobic conditions. Readings (concentrations) from control sample which is blank media with iron (III) citrate and no culture was subtracted from the NDH-2 and uninduced culture reading to rule out the chemical reduction of iron citrate. The maximum amount of ferric citrate reduced by induced culture was 0.94 mM compared to uninduced culture which was 0.59 mM by the end of the 6th day which was ~1.6 fold higher than the uninduced (Fig 3.17). In addition to the total ferric citrate reduced by the end of day 6, it was also observed that the rate of reduction was same in both induced and uninduced cultures till 3rd day while the induced culture almost continued at the same rates until day 6, uninduced cultures showed reduced reduction rates after day 3. The enhanced reduction of ferric form to ferrous form can signify the enhanced extracellular electron transport as a result of NDH-2 expression.

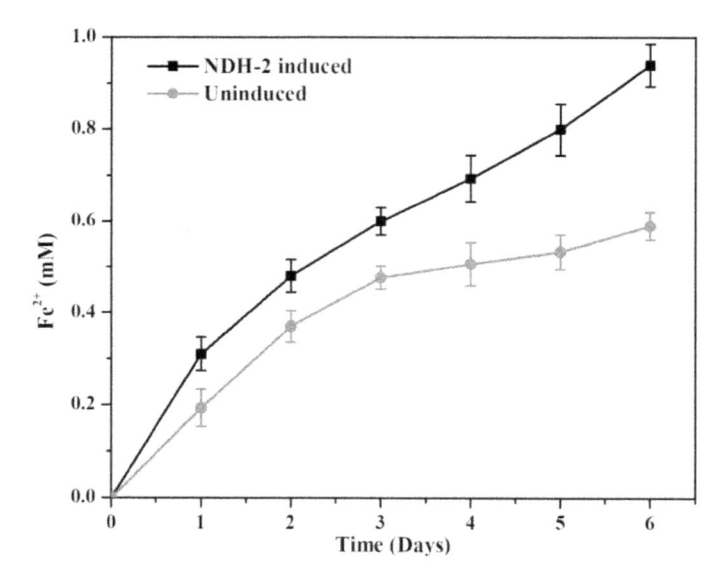

Fig 3.17: Iron reduction by NDH-2 and uninduced cells using Ferrozine assay with respect to time.

3.3.9 Spectro-electrochemical studies on purified NDH-2

Spectro-electrochemistry couples electrochemistry and spectroscopic approaches, and thus enables us to follow a specific molecular response as a consequence of a chosen electrode potential. Spectro-electrochemistry can be a useful technique to study the midpoint potentials of a redox protein. The UV/VIS spectra of a protein are often specific for the redox state of a cofactor, e.g., hemes, FeS clusters or flavins. UV visible spectra of electrochemically reduced protein, when oxidized slowly at a scan rate of 1 $mV.s^{-1}$ has resulted in a distinct peak at around ~520 nm. Peak height has increased with oxidation (increasing potential) which signifies redox reaction between NDH-2 and working electrode involving groups having absorption maxima at ~520 nm (Fig 3.18).

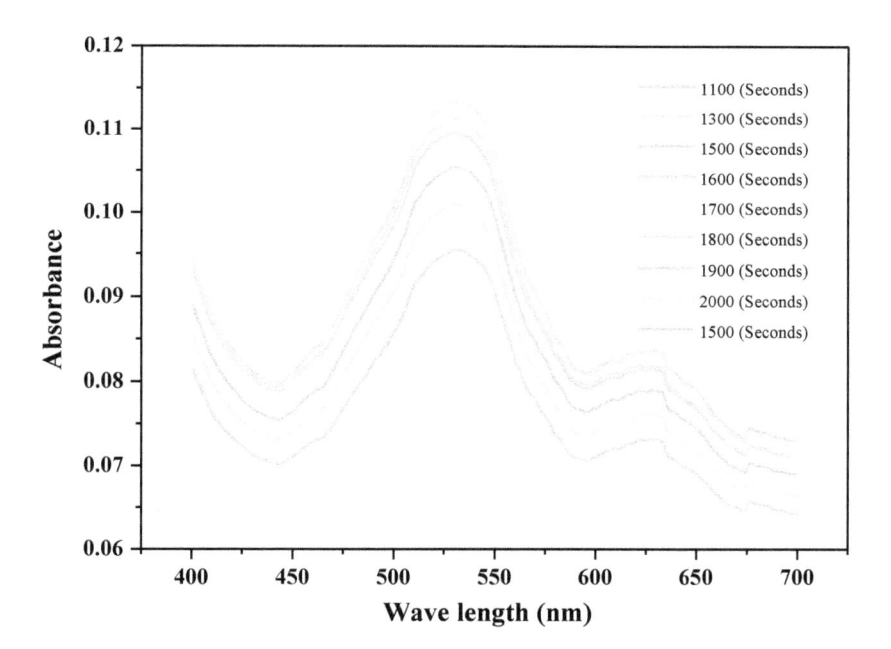

Fig 3.18: UV visible spectra of reduced NDH-2 when slowly oxidized using Potentiostat at a scan rate of 1 $mV.s^{-1}$.

This peak was not observed when NDH-2 was exposed to UV visible light without applying potential (Fig 3.10) which signifies the involvement of redox reactions in NDH-2 when subjected to electrochemical signals. Further studies are needed to be performed in order to completely understand the reaction kinetics and midpoint potentials of the enzyme.

3.4 Conclusions

This study has demonstrated the ability of NADH dehydrogenase II to enhance the metabolic flux and electron flux by increasing the NADH oxidation in the cell. Higher levels of NAD^+ and total NAD pool in the NDH-2 induced cells can be an indication of the increased electron transfer into the electron transport chain and increased catabolic activity in the cell. '*ndh*' gene expression has led to nearly threefold increase in current production 0.6 µA to 1.9 µA and ~1.6 fold increase in iron (III) reduction rates from 0.59 mM to 0.94 mM when compared to uninduced culture, which shows that the *ndh* gene overexpression has led to significant increase in extracellular electron transport. Activity of NDH-2 also seems to have positive effect on biofilm formation on anode which can significantly increase the extracellular electron transport. This study portrays the effect of NDH-2 activity on bioelectrogenesis and highlights the importance of NAD pool and activities of enzymes like NADH dehydrogenases on overall bioelectrogenesis.

CHAPTER 4
Development of Electrogenic Consortia by Selective Enrichment Strategies and Application for Real-field Scenario

Chapter 4   Development of Electrogenic Consortia by Selective Enrichment

4.1 Introduction

Indiscriminate use of non-renewable energy sources like petroleum and coal has resulted in rapid rise of pollution and earth surface temperature (global warming). In addition to this, wastewater, which is being produced in huge amount every day is discharged into fresh water streams without proper treatment. In the wake of these problems, technologies like microbial fuel cells (MFC) are being developed which can carry out both electricity generation and wastewater treatment in one system and its ability to integrate into existing wastewater treatment plants makes it practically more viable (Venkata Mohan et al., 2014a) In a MFC, electrogenic bacteria, which are considered as biocatalyst utilize the available organic substrate present in the anode chamber liberating electrons on to anode, which will be reduced at cathode in prescence of oxygen to generate electricity (Venkata Mohan et al., 2014b). Though pure strains are known to produce high power densities, mixed cultures provide better practicality for various reasons like ease of handling, avoiding sterilization of the reactor and feed. Besides, mixed culture contains diverse microorganisms which can utilize wide range of organic substrates present in various wastewaters, resulting in complete degradation (Nevin et al., 2008). Despite its advantages, untreated mixed cultures as inoculum in microbial fuel cells exhibit less substrate to power conversion efficiency because of the growth of non-electrogenic bacteria belonging to other group of bacteria like fermenters, acetogens, and methanogens (Jung and regan., 2011). These non-electrogenic bacteria can compete with the electrogenic bacteria for substrate and can also form biofilm on electrode blocking the anode surface area available for the electrogenic bacteria, which can drastically reduce the coulombic efficiency of the system. Moreover, electrogenic bacteria can be outcompeted by several groups of bacteria like methanogens. Hydrogenotropic methanogens can also utilize the reducing equivalents produced by electrogenic bacteria for the production of methane and also other metabolites effectively reducing the efficiency of MFC (Burak and Paul., 2008).

Anode potential can be considered as the electromotive force that drives electrons from bacteria to the anode, and it is regarded as a measure of electron affinity of the electrode. Anode potential can induce growth of the biofilm by acting as a sink (final electron acceptor) for electrons for the bacteria.It has been observed from several studies that the anodic community composition and biomass composition changes according to the external resistor or applied potential (Pham et al., 2008; Wang et al., 2010) which can be exploited for the enrichment of electrogenic bacteria in MES. Anode potential directly controls the biofilm formation on the anode there by influencing their overall performance and efficiency of MES. Anode potential can be controlled either extrinsically using a potentiostat or by using an external resistance. Potentiostat polarize the electrode by varying the electrical energy input, while external resistances simply regulate anode potential by using the *in situ* potential developed without external energy input.

Enrichment of inoculum using applied potential can decrease the startup time of the reactor because of the rapid formation of electrogenic biofilm when potential is poised (Gimkiewicz and Harnisch, 2013). Applied potential also favors the growth of exoelectrogen species such as *Geobacter, Shewanella, Pseudomonas* etc. in the absence of oxygen and was confirmed by several microbial diversity studies of MES (Gimkiewicz and Harnisch, 2013). By maintaining relatively high anode potential (370 mV; vs. SHE electrode) using graphite rod electrodes, thin biofilms with high microbial diversity were formed. While lower anode potential values of 150, 90, and 20 mV have produced thick biofilms with a good fraction of bacteria belonging to genus *Geobacter* (Torres et al., 2009). Prolonged application of positive anode potentials (+0.5/+1/+2 V) to MES showed wide diversity in bacterial enrichment in microbial fuel cells (Kannaiah Goud and Venkata Mohan, 2013). Among the three positive potentials, þ1V condition has resulted in maximum performance with enrichment of EAB belonging to Proteobacteria. Computational models have predicted that increased external resistance in MFC can increase the biomass and decrease the current generated at the anode (Picioreanu et al., 2008). It should also be noted that the cell configuration can also influence the current densities generated (Vázquez-Larios et al., 2011). In single-chambered MFC, high current densities were observed when high positive potentials were used, whereas in the case of double-chambered MFC, negative potential has shown to produce more current densities.

Electrogenic bacteria such as *Geobacter, Shewanella*, and *Acidithiobacillus* genus etc., are capable of naturally reducing/oxidizing the insoluble metal oxides (iron, manganese, uranium, etc.) when no other electron acceptors or donors are available in the environment (Lovley, 2012). Whereas, non-EAB fail to sustain active growth when metal oxides are provided as only electron acceptors. This property of electroactive bacteria can be exploited for their enrichment. Preincubation of inoculum with poorly crystalline iron oxide resulted in enriched exoelectrogens, which can also reduce iron oxide faster (Wang et al., 2010). When the enriched culture was used as an inoculum for MES, it also resulted in higher power density and reduction in the startup phase.

Shear rate is the rate of change in velocity at which one layer of fluid passes over an adjacent layer, which plays an important role in biofilm formation, especially when operated in continuous mode. Higher shear rates can result in stronger aggregation and attachment of microbes when the shear rate is maintained below the tensile strength level (Pham et al., 2008). This results in formation of thicker biofilms, thereby transferring electrons by direct electron transport via cytochromes or nanowires, in addition to using redox mediators. High shear rate also eliminates the rapid pH fluctuation and high mass transfer (nutrients), which will ultimatelyresult in the stability of the system and in the increase of electricity generation. In the case of shear rates above critical tensile strength, microbes tend to detach from the electrode, resulting in weaker biofilm formation. The bacterial density of biofilm from the high shear enriched MES will be substantially higher than that of the low shear mES.

4.2 Selective enrichment strategies for electrogenic consortia: effect of heat and Iodopropane pretreatment

As the performance of MES can be enhanced by using optimum inoculum, in this objective we sought to evaluate the heat and Iodopropane pretreatment strategies for selective enrichment of electroactive bacteria and understand its effect on microbial community structure (Fig 4.1).

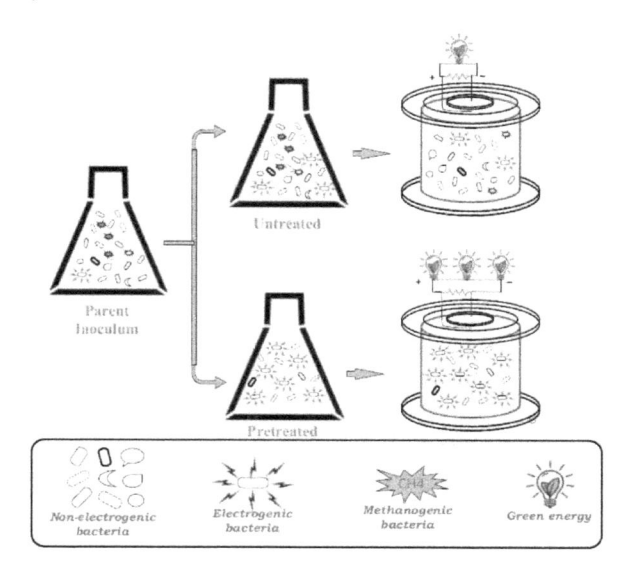

Fig 4.1: Schematic diagram of pretreatment strategy used for enrichment of electroactive bacteria.

Specific methanogenic inhibitors like Iodopropane and 2-bromoethanlsulfonic acid (BESA) were widely used in the biohydrogen reactors to prevent the growth of both acetoclastic and hydrogenotropic methanogens which utilize acetate and hydrogen, respectively (Venkata Mohan et al., 2008; Ghimire et al., 2015, Angenent et al., 2004). It was also considered that the methanogens are the main competitors for the electrogenic bacteria in MFC operating at anaerobic conditions decreasing the overall power output. Iodopropane and BESA pretreatment strategy can also be utilized in MFC as they will be inhibiting the specific pathways responsible for production of methane via separate mechanisms (Kenealy and Zeikus., 1981; Bouwer and McCarty., 1983), thereby preventing the growth of methanogens without effecting the growth of electrogenic bacteria and without changing the composition of the bacterial diversity. Iodopropane (Propyl Iodide) inhibits the metabolism involving the corronoids ([Co]E_2) as methyl carrier. In methanogens, methyl corronoid is the central molecule for both the anabolic and catabolic pathways like acetyl coA synthesis, methanogenesis etc. In the prescence of Iodopropane [Co]E_2 binds to propyl group instead of methyl group thus inhibiting the pathways requiring the methyl groups and thereby growth itself (Simpson and Whitman., 2012). Whereas nonspecific pretreatment methods like heat can result in elimination of many species of microorganisms except endosporulating bacteria and thermophilic bacteria. Heat pretreatment also inhibits the growth of methanogenic bacteria as they cannot survive the heat pretreatment.

4.2.1 Experimental

4.2.1.1 Biocatalyst and Selective enrichment methods

Anaerobic granular sludge acquired from a full-scale local municipal sewage treatment plant was used as original inoculum for the experiment. Granular sludge was disaggregated and separated from debris by passing through sieve. The pretreatment methods applied to the inoculum were Iodopropane and heat. Iodopropane pretreatment was applied by adding 2-Iodopropane solution to 150 ml of sludge at a final concentration of 50 mM. Once Iodopropane solution was added to the sludge, it was mixed thoroughly and incubated at room temperature for 24 hours in anaerobic conditions before adding to the reactor. Iodopropane is considered as an inhibitor of corrinoid-dependent metabolism. It binds to corrinoid enzymes and inhibits their activity (Kenealy and Zeikus., 1981). Heat pretreatment was performed by heating the sludge at 80°C for one hour. Both Iodopropane and heat pretreatments were performed at room temperature and atmospheric pressure.

4.2.1.2 Experimental design and operation

To know the efficiency of various pretreatment strategies over the untreated inoculum, three dimensionally similar microbial fuel cells (MFCs) were constructed using Perspex material. Total volume of each reactor was 700 ml with 550 ml as working volume. All the three systems were single chambered and fixed with non-catalyzed graphite plate electrodes of dimensions 5 x 5 x 1 cm. Anode was submerged in anolyte (designed synthetic waste water) while the cathode was exposed to atmosphere (air cathode). Nafion membrane acquired from Sigma Aldrich was used a proton exchange membrane and was sandwiched between anode and cathode which will make sure that only protons are transferred from anode to cathode via membrane. Redox mediators were not used in this study. All the three systems were operated in fed-batch mode in anaerobic conditions at room temperature. Magnetic stirrer (100 rpm) was employed to mix the anolyte during operation to prevent settling of the inoculum. All three systems were operated with designed synthetic waste water (DSW) with glucose as carbon source [glucose – 3 g/l; NH_4Cl – 500 mg/l, $MnCl_2$– 15 mg/l, KH_2PO_4 – 250 mg/l, K_2HPO_4 – 250 mg/l, $CuCl_2$ – 10.5mg/l, $MgCl_2$ – 300 mg/l, $CoCl_2$ – 25 mg/l, $ZnCl_2$ – 11.5 mg/l, $CaCl_2$– 5 mg/l, $FeCl_2$ – 25 mg/l, $NiCl_2$ – 16 mg/l] at a pH of 7. pH of the feed was adjusted using either 0.1N HCl or 0.1N NaOH. pH of feed was only set as desired and change in pH was monitored with time. Each cycle (hydraulic retention time) was run for 48 hours, where exhausted feed was exchanged with fresh feed. All the experiments done here were performed thrice and the average values were presented in the manuscript.

4.2.1.3 Analysis

Electrogenic parameters viz., current and voltage were continuously monitored across an external resistance of 100 Ω connected between anode and cathode using a digital multimeter. During start-up cycles, fluctuations in current and voltage profiles were observed which were subsequently stabilized. Once the stabilization phase is reached, polarization measurements were performed. To generate the polarization curves, voltage was measured using a series of resistors ranging from 50 Ω to 30 kΩ. Biochemical parameters like chemical oxygen demand (COD), volatile fatty acids (VFAs) and pH were analyzed from time to time according to the standard methods described in APHA, 1998. COD was performed at the end of each cycle by using closed reflux method and COD removal efficiency was calculated accordingly. Bio-electrochemical characteristics of the MFCs were studied with a potentiostat–galvanostat system (Bio-logic instruments- VSP 1, France) and the obtained data was analyzed using EC lab software v10.4. All the bio-electrochemical experiments were performed by considering anode and cathode as working and counter electrodes, respectively. Ag/AgCl; KCl (3.5M) was used as reference electrode for all the experiments. Cyclic voltammograms were acquired by ramping potential between -0.5 to 0.5 V on working electrode (anode) at 20, 30, 50 and 100 mv.s^{-1} scan rate. Chronoamperometry was performed to study the ability of the system to release electrons under applied potential. It was done by applying a potential of 0.5 V on working electrode for 30 minutes and sustainable current generation by the system was analyzed. Coulombic efficiency (CE) was calculated according to the formula described by Suresh babu et al., 2014.

$$CE = \frac{M \int_0^t i\, dt}{Fbv_{An}\Delta COD} \times 100 \qquad (4.1)$$

Where i is the current (mA), F is the Faraday's constant (96485.33 C/mol), M is the molecular mass of oxygen (32 g), v_{An} is the volume of liquid in the anode compartment, b is the number of electrons exchanged per mole of oxygen (4 electrons), and ΔCOD is the change in COD over a period of time (t).

4.2.3 Results and discussion
4.2.3.1 Bioelectrogenesis- pretreated Vs untreated

During the start up of the MFCs, it was observed that the current generation from all the MFCs was fluctuating and showed a similar pattern. From the Fig 4.2, it can be understood that the power output of the reactors was influenced by the pretreatment of the inoculum. Highest power density (PD) of 180.5 mW.m^{-2} was observed in the case Iodopropane pretreated MFC at a current density (CD) of 398.6 mA.m^{-2} during the 5th cycle of operation. While the heat pretreated MFC, and untreated MFC (control) have shown highest power densities of 128.26 mW/ m^2 and 92.08 mW/ m^2, respectively at a current density of 345.71 mA.m^{-2} and 284.84 mA.m^{-2} which is 13.5 % and 28.7 % less than Iodopropane pretreated MFC. Though initial fluctuations were observed with respect to power output in both the pretreated MFCs, it can be seen from the Fig 4.2 that the power output has increased substantially from 4th cycle of operation, which was more prominent in case of Iodopropane when compared to heat.

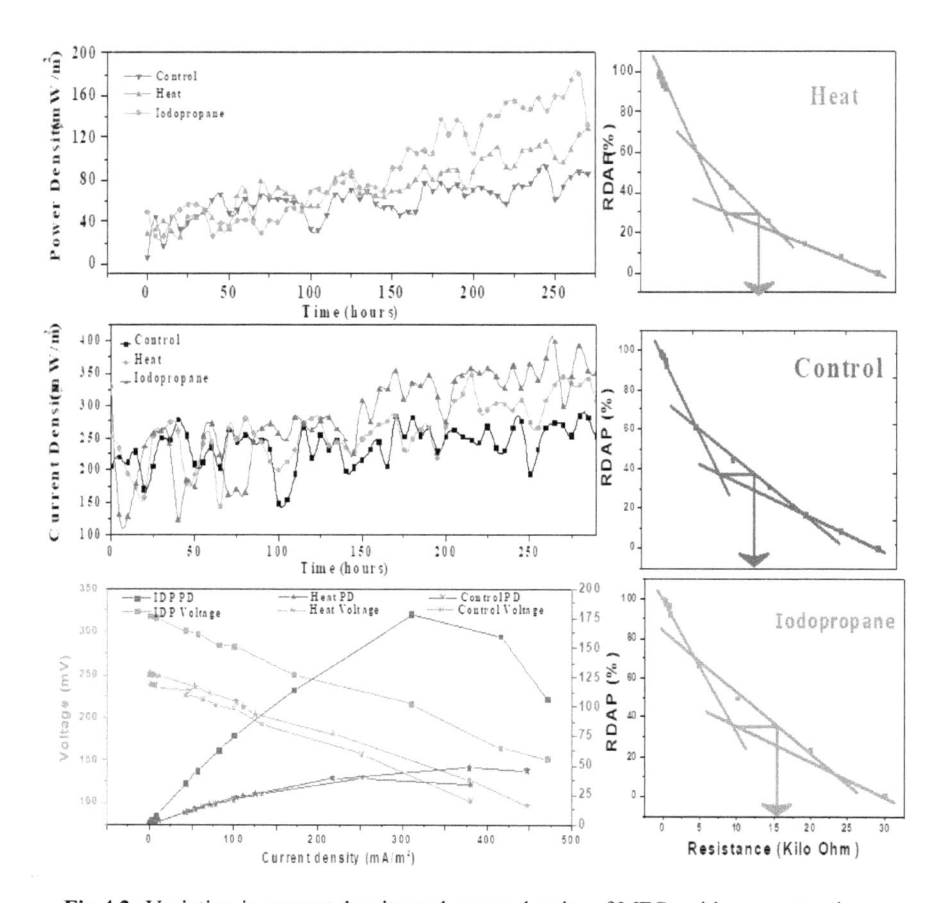

Fig 4.2: Variation in current density and power density of MFCs with respect to time.

Whereas in the case of control MFC, constant but low power output was observed (CD ~ 250 mA/m^2; PD ~ 80 mW/m2) during the entire experimental period. In case of both the pretreatments, it was observed that the stabilization period and time required for the biofilm formation has decreased considerably when compared to the untreated control. This also indicated that the pretreatment has also enhanced the biofilm formation on electrode leading to stable electrogenesis. Higher power output can also be correlated well with the microbial diversity studies which have shown that pretreatment eliminated most of the non-electrogenic bacteria compared to untreated control and crude (parent) inoculum.

4.2.3.2 Polarization behaviour

Polarization curve gives the information on the cell design point and various losses (ohmic, concentration and mass transfer) occurring in the system (Venkata Mohan et al., 2014b; Zhao et al., 2009). They were plotted with current density against potential and power density at different resistances (100 Ω to 30 KΩ) to visualize the maximum power density and current density with respect to pretreatment. Polarization curves were plotted when reactors showed the stable performance (Fig 4.2). Iodopropane has shown cell design point at 450 Ω indicating that the system can be operated even at a load of 450 Ω. This is followed by heat (300 Ω) and control (250 Ω). Cell design point (CDP) clearly showed that the Iodopropane MFC is more robust and can be operated at higher loads for producing stable current. Relative decrese in anode potential (RDAP) is performed to know the effective resistance at which MFC can produce electricity sustainably. It was performed by measuring cell potential and current with respect to given resistance by stepwise decrease in the resistance from 30 k Ω to 50Ω. (Menicucci et al., 2006). Iodopropane showed highest resistance (16 kΩ) for production of sustainable current followed by heat (14 kΩ) and control (12 kΩ).

4.2.3.3 Coulombic efficiency (CE) and substrate degradation

COD removal efficiency of a system gives the information on ability of inoculum to degrade the available organic substrate whereas CE can give a good understanding of efficiency of MFC to harvest electricity from the substrate utilized by the inoculum because not all the substrate utilized by bacteria is converted to electricity. Coulombic efficiency (Eq 4.1) can be defined as the number of coulombs recovered from a system as electrical current compared to the theoretical maximum number of coulombs recoverable from the organic substrate added to the system (Ashley Franks and Nevin., 2010). Highest CE of 6.188 % with 52% of COD removal was observed in case of iodopropane MFC (Fig 4.3). Comparatively heat MFC showed 4.75 % of CE with COD removal of 51%. While the untreated control showed high amount of COD removal i.e., 66% with CE of 2.89 %. Though the CE of all the MFCs was less than what was reported for pure cultures (Bond et al., 2002), it should be noted that the MFCs were run with mixed culture as inoculum and non-catalyzed graphite as electrodes which have high losses but have more practicality. It should also be noted that the CE of untreated control was 53% less than iodopropane MFC and 23% less than heat MFC depicting the efficiency of pretreatment of inoculum on exoelectrogenesis. The results of this experiment have shown the positive influence of pretreatment on substrate to electricity conversion efficiency which is in accordance with the assumption that both Iodopropane and heat pretreatment can eliminate competitors for electrogenic bacteria. Though the COD removal efficiency is less in Iodopropane pretreated MFC compared to control, it showed higher CE when compared to the latter. This might be because of the shift of microbial diversity of Iodopropane MFC due to elimination of Methanogens. It is because of this, Iodopropane pretreated MFC is now able to convert the utilized substrate to coulombs more effectively than untreated MFC. Moreover, untreated inoculum which houses more diverse bacterial species including non-electrogens can utilize substrate for other metabolic reactions (non-electrogenic metabolism fermentation, methanognesis, solventogenesis, etc).

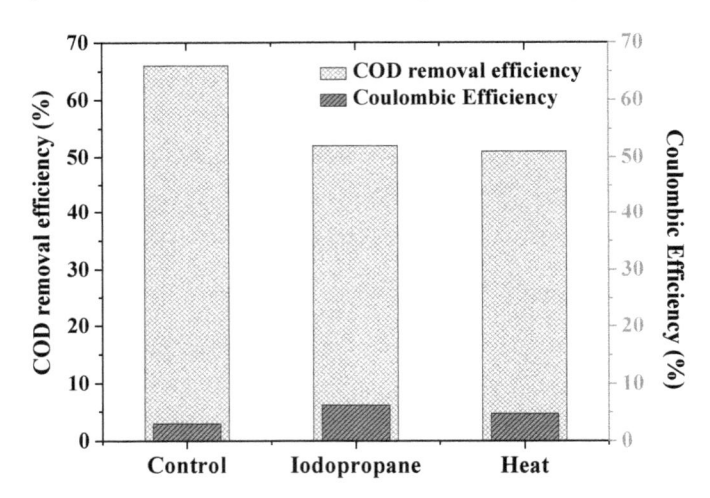

Fig 4.3: Coulombic efficiency of Iodopropane and heat pretreated MFCs with respect to untreated MFC. Untreated MFC has shown highest COD removal but low CE indicating low substrate to power conversion

Whereas in case of heat MFC, pretreatment eliminated most of the bacteria while enriching bacteria which can resist stress (like *Bacilli* and *Clostridia*) along with bacteria like *Xanthomonas, Pseudomonas* etc. From the previous studies it can be found that species belonging to genus *Bacillus, Clostridia* and *Pseudomonas* can perform exoelectrogenesis when required (Goud and Venkata Mohan., 2013). Similar studies of Venkata Mohan and co-workers with biohydrogen production experiments have indicated that the pretreatment methods like Iodopropane and heat can eliminate the methanogens which are principle competitors for the substrate and increase the production of biohydrogen (Venkata Mohan et al., 2008). There are mainly two reasons for which methanogens can be considered as main competitors in a MFC. They can deprive electrogens of available substrate for their growth and production of electricity while in other way they can utilize the protons produced by exoelectrogens in anode chamber to synthesize methane thereby hampering the electricity production and reducing the overall CE (Ashley Franks and Nevin., 2010). These results also have correlated well with the results of bioelectrogenesis where Iodopropane MFC which showed higher power output (IDP- 180.55 mW/ m^2; heat- 128.26 mW/ m^2 and control 92.08 mW/ m^2) compared to other MFCs studied. This shows that the Iodopropane has effectively inhibited the growth of competitors like Methanogens and thereby increased the amount of substrate available for exoelectrogens.

4.2.3.4 Cyclic voltammetry

Cyclic Voltammetry has been regularly used to study the electrochemical reactions at the bacterial and electrode interface. They are also used to understand the electron transport process in biofilms formed on electrode (Modestra and Venkata Mohan., 2014).

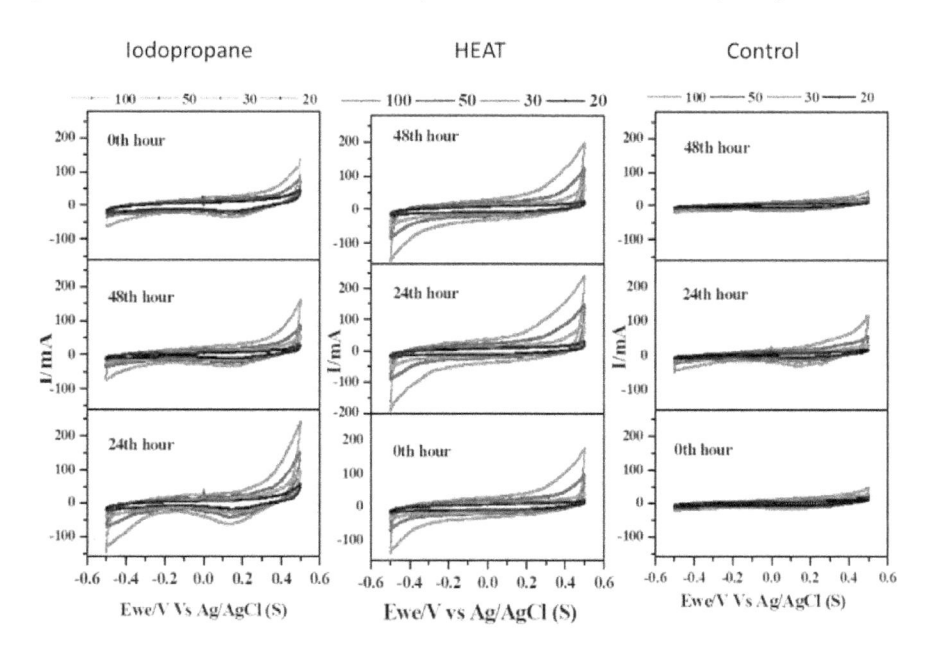

Fig 4.4: Cyclic voltammograms of Iodopropane, heat and untreated MFC during various hours of reactor operation.

As all the three MFCs operated in this study have similar operation parameters and reactor configurations except the inoculum, any variations in the cyclic voltammograms can be directly attributed to pretreatment and its effect on the biofilm formation on anode. Figure 4.4 shows the CV obtained from three MFCs operated in the experiment. CVs were taken during the stable phase of the reactor at various scan rates (20, 30, 50 and 100 mV/s). Cyclic voltammograms depicted variation between pretreated and untreated MFCs with Iodopropane showing best performance. All the three MFCs depicted high oxidation currents than reducing currents at all the time periods of a cycle at 30 mV/s. While 24th hour operation showed highest redox current. Among the various pretreatments, Iodopropane pretreated MFC showed highest currents (24th hour: oxidation, 88.35 mA; reduction, -56.67 mA) at 24th hour of operation compared to heat (24th hour: oxidation, 35.67 mA; reduction, -18.69 mA) and control (24th hour: oxidation, 32.09 mA; reduction, -15.18 mA). This can be very well correlated with the power densities of the MFCs, where Iodopropane showed highest power density followed by heat and control. Several peaks were also observed in case of Iodopropane (-97.52 mV and -364.3 mV) and control (-364.3 mV) reactors when searched for redox peaks using the peak analysis. Peaks identified in the Iodopropane can be attributed to α-ketoglutare/isocitrate redox couple (-375.9 mV) and glutathione/reduced glutathione couple (-97.52 mV). In case of control, identified peaks can be attributed to acetoacetate/hydroxybutyrate couple (-964.3 mV). Identification of peaks in CV indicates that there might be involvement of mediators in the system which enhances electron transfer efficiency. High oxidation current compared to reduction currents in Iodopropane MFC also indicates that the substrate present was oxidized effectively to liberate electrons on to anode. While in case of untreated MFC only a slight difference in oxidation and reduction currents was observed which indicates that, there might be fermentative pathways operating parallel to electrogenic pathways in untreated MFC.

4.2.3.5 Chronoamperometry

Chronoamperometry enumerates the ability of the bacteria (MFC) to release electrons onto the anode when a particular potential is applied onto anode. CA was performed at stabilized phase of MFC operation and systems ability to produce sustainable current was analyzed (Chandrasekhar and Venkata Mohan, 2012). A potential of 0.5 V was poised on to anode for half an hour. Initially a drop in current will be observed which will be followed by stable current generation. Out of the three MFCs, Iodopropane produced higher sustainable current and stabilized at 2.4 mA (Fig 4.5). While the heat pretreatment showed an initial drop in current production followed by increase in current which stabilized at 2.5 mA. Control showed lower sustainable currents 0.7 mA which is very low compared to pretreated MFCs. This indicates that the enriched inoculum in Iodopropane has more efficiency to liberate electrons than the other two systems which can be correlated well with the other analysis. This might also be a result of rapid biofilm formation in case of Iodopropane and heat MFC.

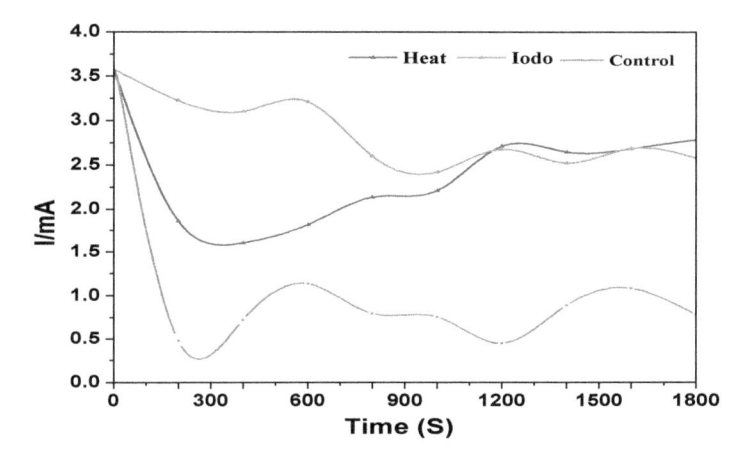

Fig 4.5: Chronoamperometry curves depicting the sustainable current. CA was performed at an applied potential of 0.5 V for 30 minutes. Iodopropane showed highest sustainable current generation.

4.2.3.6 pH and VFA

pH and volatile fatty acids give an understanding of metabolic state of the bacteria. pH values were taken using a digital pH meter and VFA was estimated using titration method according to APHA protocols. All the reactors were adjusted to an initial pH of 6 and then the change in pH with respect to time was studied (Goud and Venkata Mohan, 2012). Iodopropane MFC showed least drop in pH (4.7) with a corresponding VFA of 2650 mg/L compared to heat MFC (4.5; VFA, 2500 mg/L) and untreated MFC (4.0; VFA, 3400 mg/L) (Fig 6). pH changes of media generally indicate the shift of bacterial metabolism. Gradual drop in pH indicates the accumulation of short chain or volatile fatty acids like acetic acid, propionic acid in the media which is the result of non-electrogenic metabolism (fermentation). In addition to this, accumulation of metabolic acids can also result in inhibition of bacterial growth as most of the strains are not acid tolerant and drop in MFC performance (Russell and Diez-Gonzalez., 1998; Puig et al., 2010). This can reduce the overall process efficiency. When the pH and VFA accumulation results were correlated to CE, it can be clearly seen that the most of

Chapter 4 Development of Electrogenic Consortia by Selective Enrichment

utilized substrate in untreated MFC was being diverted to production of metabolic acids (non-electrogenic metabolism).

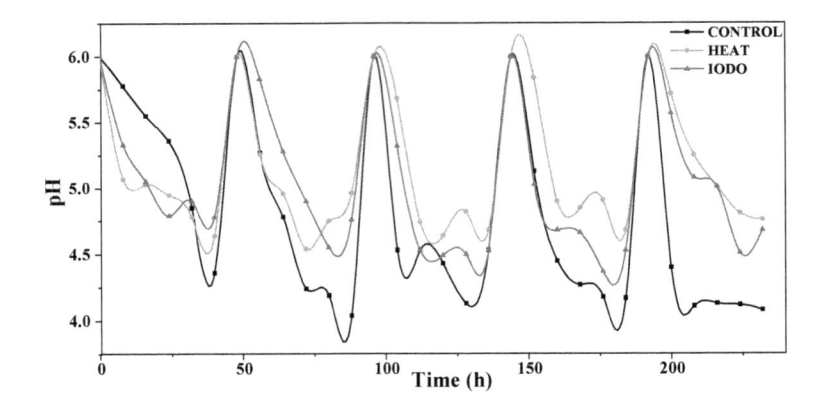

Fig 4.6: Feed with pH 6 was fed to the reactor initially and change in pH was monitored.

Presence of metabolic acids also indicates that the substrate is not completely utilized there by resulting in loss of energy in the form of VFA (Fig 4.7). While in case of Iodopropane and heat MFC, less VFA production was observed which correlated well with higher CE and might be because of the good substrate to current conversion efficiency. Selection of electrogenic bacteria by pretreatment might be a reason for less VFA production.

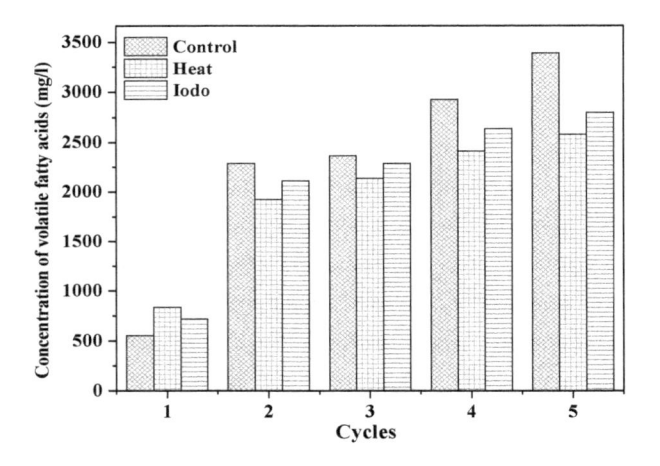

Fig 4.7: Production of metabolic acids or volatile fatty acids during MFC operation in pretreated and untreated MFCs.

4.2.4 Conclusions

Microbial fuel cells are considered as of one the emerging technologies which can integrate the process of wastewater treatment and electricity generation. In this study, pretreatment was used as strategy to enrich the exoelectorgenic bacteria. Increased power output and CE was observed in Iodopropane and heat preteated MFC when compared to the unreated MFC. Microbial community analysis has shown the domination of electrogenic bacteria in pretreated MFCs, while untreated control MFC has shown more diverse community. Overall, this work has illustrated the efficiency of pretreatment to enrich the exoelectrogenic bacteria in MFC.

4.3 Selective enrichment strategies for electrogenic consortia: Effect of acid and BESA pretreatment

The aforementioned studies indicated that the pretreatment can be used as viable option for enrichment of electroactive bacteria, so in this study we sought to investigate the additional pretreatment strategies acid and BESA for pretreatment of inoculum.

4.3.1 Experimental-Biocatalyst pretreatment

For acid pretreatment, parent inoculum was adjusted to pH 3 with concentrated phosphoric acid and the culture was incubated at room temperature for 24 hours before inoculating into the MFC reactor. For BESA pretreatment, parent inoculum was mixed with BESA and incubated for 24 hours at room temperature.

For other experimental methods refer to experimental section of 4.2.1

4.3.2 Results and discussion

Three MFCs were operated simultaneously to assess the performance of the fuel cell in terms of power generation using design wastewater treatment (DSW) as feed and glucose as carbon substrate. Anaerobic mixed consortia was pre-treated separately using acid (acid-shock), and BESA was used as inoculum for each of the fuel cells. Performances of the pre-treated fuel cells were compared with an untreated (control) parent culture. The performance of the fuel cell was studied continuously for several cycles with each cycle spanning 48 hours. Fuel cells were operated at a pH of 6.0 with glucose as substrate (carbon source) with loading of 6 g COD/l. The reactor performance was evaluated in terms of different parameters like bio-analytical (OCV, Current, pH and ORP), bio-process (Glucose, VFA and COD), bioelectrochemical (CV). Electrogenesis measurements made were represented as open circuit voltage, current generation (100 ohms) with resister and without resistor.

4.3.2.1 Bioelectrogenesis

All the pretreatment strategies have resulted in enhancement of bio-electrogenic performance of the systems. This might be due to the enrichment of electrochemically active bacteria or elimination of non-electrogenic bacteria or both. Experimental data has visualized variation in the process performance and bio-electricity production behavior with the function of the pretreatment method applied. Out of all the pretreatment strategies, acid pretreatment has showed highest electrogenic performance followed by BESA and control MFC (Fig 4.8).

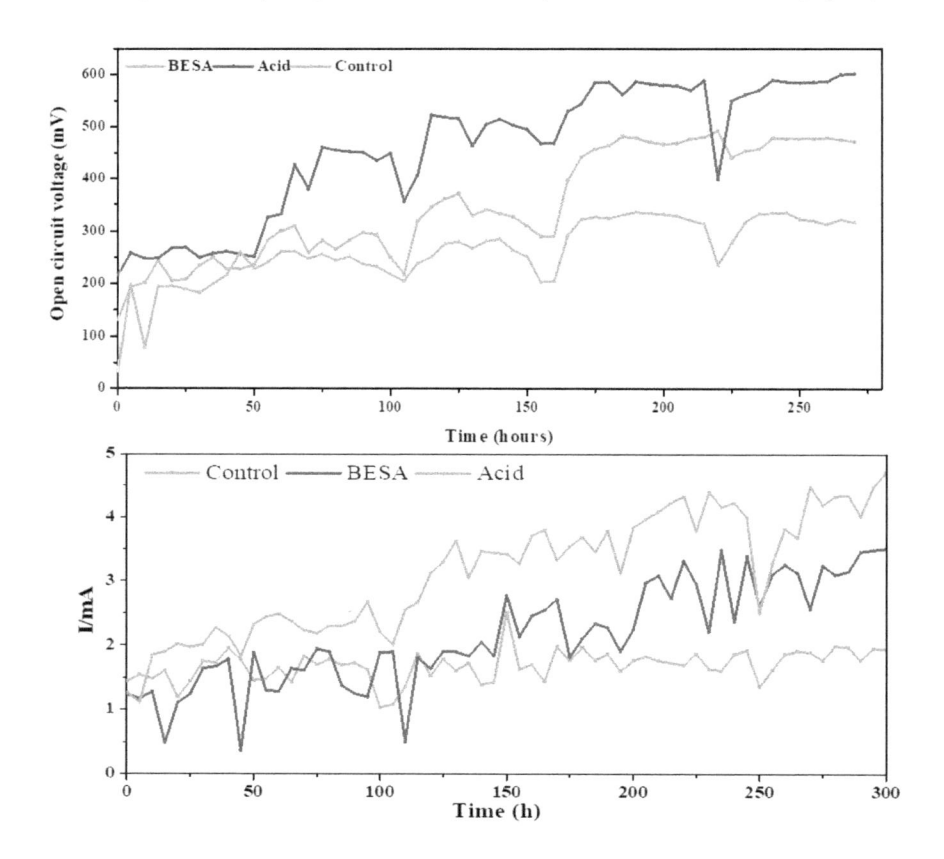

Fig 4.8: Open circuit profile and current profile with 100 Ω resistor of acid and BESA pretreated MFC with respect to control.

Though initial cycles have documented similar performance in terms of open circuit voltage (OCV), effect of pretreatment started documenting the differences with progress of time. Initial lag of performance might be due to the adaptability of the biocatalyst to the media. Acid pretreated fuel cell showed rapid rise of OCV with highest OCV (602 mV) recorded at 265th hour. OCV increased progressively with time and stabilized at ~580 mV. OCV and current production pattern could be attributed to the function of biocatalyst indicating that acid pretreatment has resulted in enrichment of biocatalyst.

Next to acid pretreatment, BESA pretreated fuel cell has showed good performance with stable voltage generation of ~450 mV. It showed highest voltage of 482 mV at 185th hour of operation. Whereas, MFC with untreated mixed culture (control MFC) (335 mV) showed lowest performance in terms of voltage generated among all the reactors operated. Current production was recorded with 100 Ω resistor and the current production pattern was recorded. From the recorded data it is clearly visible that pre-treatment of inoculum has resulted in higher current production from MFC (Fig 4.8). Maximum current was observed to be 4.3 mA for acid pretreated fuel cell at 290th hour. Acid pretreated fuel cell showed consistent increase in current production from first cycle itself. Next to acid pretreatment, BESA has showed increased effect on current production. It showed a maximum current of 3.49 mA at 235th hour of operation. Control (untreated fuel cell) has showed least performance among all the reactors (1.87 mA at 295th hour).

Untreated mixed culture (control) generally has greater bacterial diversity of bacterial population which includes both exoelectrogens and non-electrogens, whereas pretreatment of mixed culture will result in the selective enrichment of the exoelectrogens leading to less diverse but exoelectrogenic consortia. BESA which inhibits the growth of methanogenic bacteria, which are one of the main competetors for both substrate and reducing equivalents in MFC. Experimental data showed the feasibility of controlling microbial metabolic function by application of pretreatment to the native microflora. Under performance of untreated control fuel cell might be due to the prescence of non-electrogenic bacteria which compete for the organic susbstrate with electroactive microorganisms and may also utilize the reducing equivalents generated by the exo-electrogenic bacteria for production of other byproducts like metabolic acids, methane, etc. (Goud and Mohan, 2012).

Chapter 4 — *Development of Electrogenic Consortia by Selective Enrichment*

4.3.2.2 Polarization

Acid treated fuel cell (power density: 127.85 mW/m^2; current density: 597.14 mA/m^2) showed maximum power density and current density among the all the fuel cells followed by

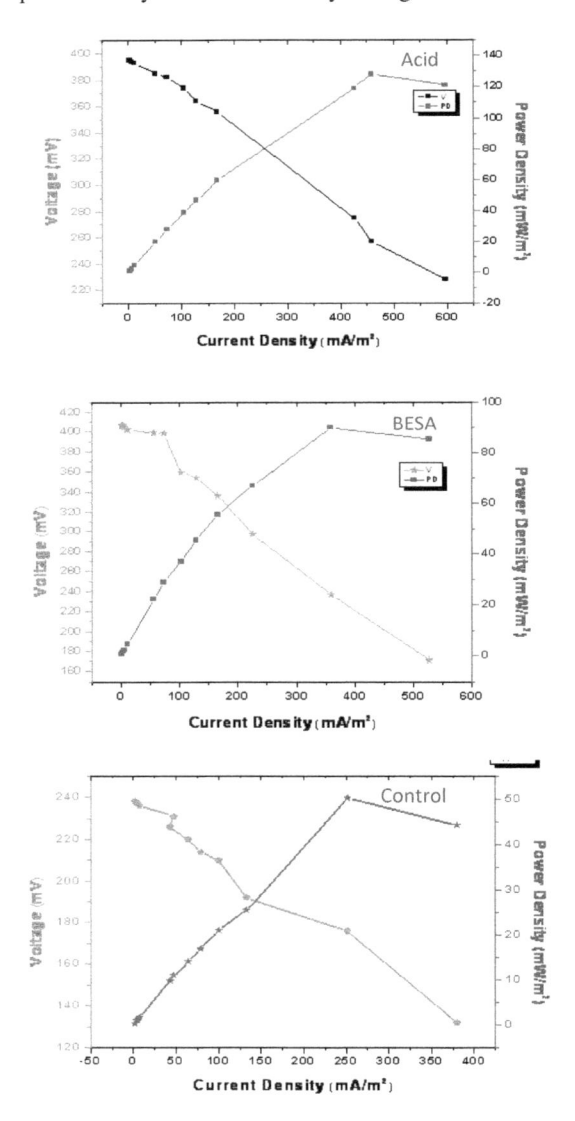

Fig 4.9: Polarization curves of all the pre-treated fuel cells with respect to untreated fuel cell.

BESA (power density: 89.98 mW/m², current density: 527.17 mA/m²), and control (power density: 50.16 mW/m², current density: 380 mA/m²) (Fig 4.9). This clearly indicates that acid pre-treatment has shown positive effect on the MFC performance. The point at which maximum power density (PD_{Max}) is observed on the polarization curve against voltage and current density is generally considered as cell design point or CDP (voltage change region) of that particular fuel cell system. Cell design points were also obtained from the polarization curve.

4.3.2.3 Columbic efficiency and substrate degration

Among all the fuel cells operated, untreated control fuel cell showed highest substrate removal (COD removal) (67 % by the end of 5th cycle). This might be attributed to diverse group of bacteria that contains wide range of bacteria which can degrade diverse group of compounds present in the anolyte.

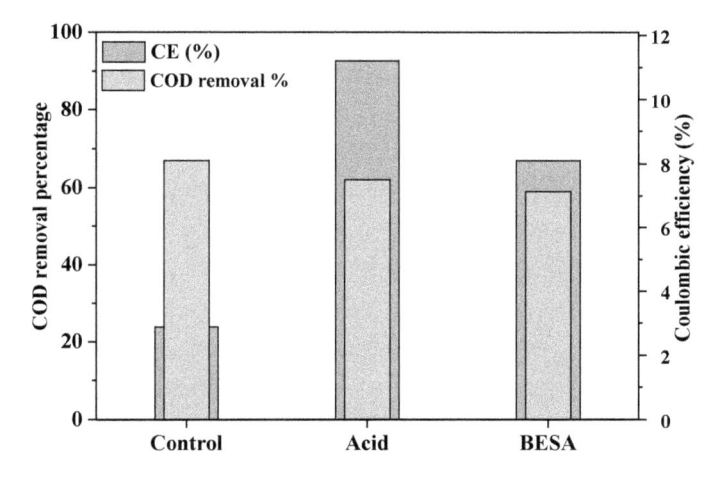

Fig 4.10: COD removal efficiency and coulombic efficiency of acid and BESA pretreated MFC with respect to untreated MFC.

To understand the substrate utilization to the power production, coulumbic efficiency for all the reactors was calculated (Fig: 4.10). Though there was more substrate degradation in control reactor, there was less performance with respect to power. This might be because, not all bacteria present in inoculum are electroactive and cannot transfer electrons to anode causing losses in the potential. Next to the control fuel cell, acid fuel cell showed good substrate removal correlated well with the high electrogenic activity of the fuel cell.

Acid pretreated reactors also showed good removal of COD showing a maximum substrate removal of (59% in 5th cycle) followed by BESA (52% COD removal percent). Highest coulumbic efficiency was seen in acid pretreated fuel cell (11.2%) which can be correlated to the current production efficiency of the system. This is followed by BESA pretreated fuel cell (7.9%) and control which has showed least coulumbic efficiency of 2.89%. Low power output in untreated control reactor can be related to the low coulumbic efficiency where most of the substrate is being used by the non-electrogenic bacteria which are part of natural flora. But in case of pretreated fuel cells, enrichments have resulted in selection of electroactive bacteria or suppression of non-electrogenic bacteria which has resulted in more power output per substrate consumed.

4.3.2.4 Cyclic Voltammetry

Electrochemical behavior of different pre-treated and untreated reactors was evaluated by using cyclic voltammetry (CV). Voltammograms (Vs Ag/AgCl (S)) were taken for all the reactors visualized marked variation in the electron discharge (Fig 4.11). CV can also be used to determine the electron stoichiometry of a system (k_{app}), the diffusion coefficient of an analyte, and the formal reduction potential etc (Modestra and Venkata Mohan, 2014). The reactors performance was assessed by considering the anode as working electrode, cathode as counter electrode and Ag/AgCl, KCl (3.5 M) as reference electrode by applying a potential ramp at various scan rates in the potential range of -0.5 to 0.5 V. All the fuel cells showed increment in redox catalytic currents during 0th hour to the 24th hour and decreased during 24th hour to 48th hour of operation showing relation with the electrogenic activity of all the reactors. All the reactors showed their metabolic activity towards oxidation rather than reduction depicting a typical performance of bacteria in a microbial fuel cell. From cyclic voltammogram it can be seen that acid pretreated fuel cell was able to discharge more reducing equivalents than BESA and control which can is evident from the high redox currents produced from acid pretreated MFC. Acid MFC was followed by BESA and control. In addition to his, several peaks were identified in both pretreated and untreated MFCs indicating the involvement of either soluble redox mediators or membrane bound protein complexes.

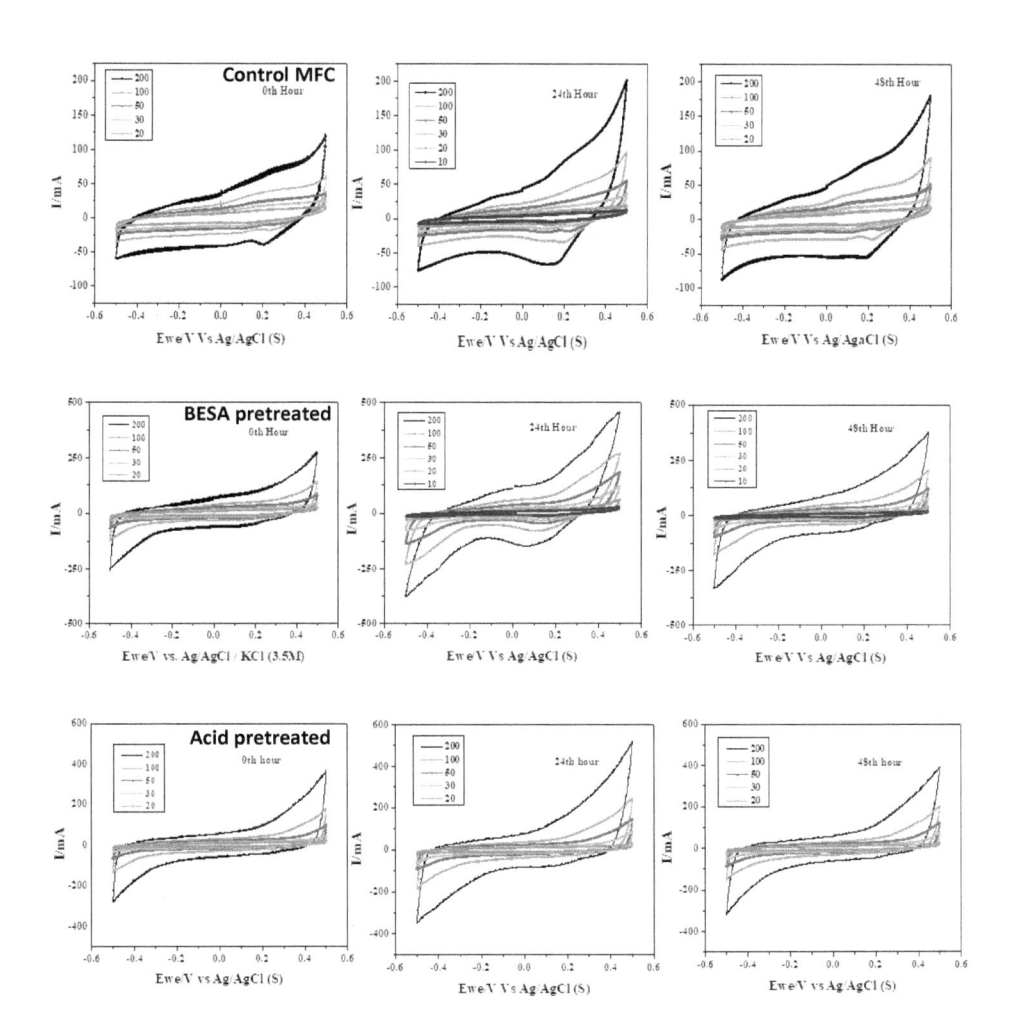

Fig 4.11: Cyclic Voltammogram profiles of acid, BESA and control MFCs at various scan rates scan rates from in the scan range of 0.5 V to – 0.5 V.

Whereas untreated control fuel cell has resulted in lowest redox currents among all the pretreatment reactors. Redox current taken at 24th hour of operation at 30 mV/s for all the reactors were as follows (Table 4.1). From the above table it can be clearly understood that untreated control fuel cell has produced least redox currents. Heat pretreated reactor has shown only marginally better performance than other pretreated reactors which can also be seen from OCV and current profiles.

Table 4.1: Redox currents obtained from cyclic voltammograms during various hours of operation at 30mV/s scan rate.

	0th hour		24th hour		48th hour	
	OC (mA)	RC (mA)	OC (mA)	RC (mA)	OC(mA)	RC (mA)
Control	17.93	-9.99	32.09	-15.18	19.66	-10.22
Acid	49.9	-22.42	148.79	-68.08	56.66	-13.96
BESA	76.63	-59.27	120.27	-77.16	45.37	-32.64

High capacitance and redox currents (oxidation current and reduction current) in acid pretreated fuel cell indicates that the electrochemically active inoculum has been enriched by pretreatment which rapidly oxidise the substrate and generate reducing equivalents which will be donated to electrode (anode). Low power generation in untreated control fuel cell indicated that the non-electrogenic bacteria like fermenting bacteria which uses reducing equivalents for production of other solvents and acids (volatile fatty acids), methanogens which uses reducing equivalents to synthesise methane etc have been enriched along with electrogenic bacteria. Cyclic voltammograms were peak searched for the involvement of redox couples identified as peaks. The obtained peak potentials (Epeak) (against Ag/Agcl,KCl(3.5 M) reference electrode) were adjusted to standard hydrogen electrode (SHE) and correlated with standard redox potentials (E$_o$ Redox) of various membrane localized metabolic reactions and electron transport chains in bacteria (Metzler, 2003). Most of the redox couples identified were as follows (Table 4.2). Involvement of several of mediators in both the pretreated fuel cells indicate that the robustness of the micobiome present in the fuel cell.

Table 4.2: Peak potentials identified in cyclic voltammograms.

Pretreatment method	Peak potentials vs SHE	Mediator identified
Acid	-67	Dehydroascorbic acid/ ascorbic acid
	92.5	Ubiquinone/ ubiquinone H$_2$
BESA	-166.25	Oxaloacetate/ Malate
	30.1	Fumarate/ succinate
	-80.1	Gluthathione/ reduced glutathione
	-375.9	α- ketoglutarate/ isocitrate
Control	-364.3	Acetoacetate/ hydroxybutyrate

More incidences of redox peaks in pretreated fuel cells might be because of the enriched electrogenic inoculum which has increased membrane bound redox protein to exchange electrons with electrode. Presence of peaks in CV also indicates that there might be an involvement of redox mediators or membrane bound redox mediators

4.3.2.5. Chrono-amperometry (CA)

Chronoamperometry was performed by poising the anode with a potential of 0.5 V for fifteen minutes and the current production pattern will be analyzed. Current produced in CA application is mainly due to the faradic processes (electron transfer based) (Chandrasekhar and Venkata Mohan, 2012) between bacteria (biocatalyst) and the electrode. Acid pretreated (3.41 mA) MFC showed highest sustainalble current among the 3 MFC systems which was followed by BESA pretreated (3.35 mA) and control MFCwhich produced lowest sustainable current of 0.7 mA (Fig 4.12). High amount of faradic current in acid and BESA pretreated MFCs may be due to the high electron holding capacity electrogenic biocatalyst which has been enriched by pretreatment.

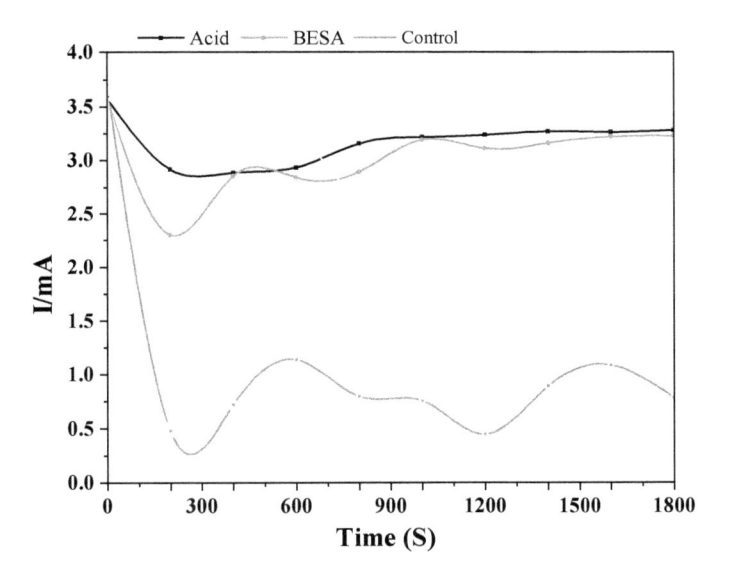

Fig 4.12: Chronoamperometric profiles of pretreated MFC compared to untreated control MFC.

Chapter 4 Development of Electrogenic Consortia by Selective Enrichment

4.3.2.6 pH and VFA variations

pH is very important factor which influences the metabolism and growth of bacteria. Previous report by Venkata mohan et al., 2007, 2008 has shown that acidic pH i.e., 6 is most favorable for the operation of MFC. This is because lower pH has certain advantages over neutral and basic pH. Acidic pH is known to favor the acidogensis generating more reducing equivalents and also simultaneously suppressing the growth of methanogens which are the main competitors for these reducing equivalents (Goud and Venkata Mohan, 2012). Methanogens generally use the available reducing equivalents to reduce CO_2 to methane (Venkata Mohan et al., 2007, 2008).

During MFC operation, initially the substrate feed was adjusted to pH 6 for all the cycles to control the growth of methanogens. The outlet pH for all of the reactors was in between 4.0 to 4.8 with untreated control reactor showing highest drop to 4.04 during 2nd cycle of operation. During operation of all the cycle except first cycle, maximum drop of pH was observed in untreated control fuel cell (cycle 2, pH 4.04; cycle 3, pH 4.17; cycle 4, pH 4.17 and cycle 4, pH 4.08). During the initial hours all the MFCs showed a pH drop from to 4.5 to 5.5 by the end of the cycle. BESA pretreated fuel cells showed a drop upto pH 4.5 during all the cycles of operation (BESA, pH 4.72) by the end of the operation. Acid pretreated fuel cell (pH 4.35) showed more pH drop when compared to BESA system. This might be due to the small amount of phosphoric acid which might have left over in the inoculum. Highest pH drop was observed in case of untreated (control) fuel cell (4.25; 24th hour). Higher drop in pH of control MFC compared to BESA and acid system indicates that there is more substrate utilization for VFAs production by non-electrogenic bacteria in control MFC than pretreated reactors. Volatile fatty acids (VFA; represented as the total amount of all acids generated during acidogenic fermentation step) concentration along with pH were also monitored during the MFC operation to evaluate the bioprocess mechanism during bioelectricity production. VFA pattern has correlated well with the pH pattern of the fuel cells operated. VFA production was always associated with conversion of organic fraction to acid intermediates in the anaerobic microenvironment with the help of specific group of bacteria called fermentors. Generally acidic intermediates enumerate changes in the metabolic pathway of the ongoing anaerobic process. A gradual raise in VFA concentration was observed during MFC operation in all the reactors. Highest amount of VFA production was observed in the case of untreated control reactor (3390 mg/L) in the cycle 5 (cycle 1; 550.7

mg/l: cycle 2; 2285 mg/L: cycle 3; 2365 mg/L: cycle 4; 2927 mg/L). Less power production in untreated fuel cell compared to treated fuel cell might be attributed to the volatile fatty acid production in untreated control fuel cell. Maximum levels of VFA produced in other MFCs is acid (3199 mg/L) and BESA (2338 mg/L).

4.3.3 Conclusions

Microbial electrochemical systems are one the emerging technologies which can be integrated the current wastewater treatment systems for simultaneous electricity generation and treatment. In this study, two new pretreatment strategies were used to see if electrogenic activity of the inoculum can be further increased. Both acid and BESA pretreatments have resulted in enrichment of electroactive bacteria. Increased power output and coulombic efficiency was observed in acid and BESA preteated MFC when compared to the untreated MFC. Microbial community analysis has shown the domination of electrogenic bacteria in pretreated MFCs, while untreated control MFC has shown more diverse community. Overall, this work has illustrated the efficiency of pretreatment to enrich the exoelectrogenic bacteria in MFC.

4.4 Metageneomic Analysis to understand functional relation between microbial dynamics with enrichment

Understanding how pretreatment effected the functioning of mixed microbial cultures towards exoelectrogenesis wasdone by usingMiSeq platform by sequencing V3-V4 region of 16s rRNA coding gene.

4.4.1 DNA extraction

For Misequencing analysis, biomass for each sample was collected at the end of the operation (Iodopropane, heat and untreated control). In addition to these samples, crude inoculum (mother inoculum) was also used for comparison of changes in microbial community. For DNA extraction, 5 ml of active biomass is collected from the reactors and the biomass was pelleted down by centrifugation at 4000 RPM for 10 minutes immediately. Pellet was washed twice with phosphate buffer to remove any exo-polysaccharides. 200 mg of this pellet was used for DNA extraction using NucleoSpin soil DNA isolation kit (Macherey-Nagel, Germany). Concentration of gDNA extracted was checked using nanodrop (Thermo Scientific, USA).

4.4.2 MiSeq analysis

Library preparation was performed by using 4ng of Nanodrop quantified DNA for amplifying V3-V4 region of 16S region with specific primers as given below

Forward: 5'- TCGTCGGCAGCGTCAGATGTGTATAAGAGACAG-3'

Reverse: 5'- GTCTCGTGGGCTCGGAGATGTGTATAAGAGACAG -3'

The primers also have a 'tag' sequence that are complementary to Illumina sequence adapter and index primers from the Nextera XT Index kit V2. This round of PCR generates single Amplicon of ~430-480 bp. The Amplified products were checked on the Agarose Gel before proceeding for Indexing PCR. In the next round of PCR (indexing PCR) Illumina sequencing adapters and dual Indexing barcodes are added to 25ng of round1 product using limited cycle PCR to give a final product of ~570-620 bp. The library was diluted (1:10) and validated for quality by running an aliquot on D1000 Tape Station (Agilent).These libraries were further pooled and cleaned using high prep beads to prepare for sequencing on Illumina platform. After the sequences were quality checked using FastQC tool based on the sequence and its quality value (ASCII characters), raw data was processed using QIIME pipeline where 16s rRNA detection and clustering, OTUs picked and BIOM files generation was done. Closed-reference OTU picking was done in this study, where the reads were clustered against a reference sequence collection and any reads which do not hit a sequence in the reference sequence collection were excluded from downstream analysis.

4.4.2 Microbial Dyanmics- heat and iodopropane enrichment

4.4.2.1 Sequence analysis and diversity estimation

Total paired end reads were around 23,20,078 to 37,33,402 for 4 samples analyzed (Table 4.3). After the pair end reads were processed to remove the adapters, barcodes, chimeras and primers the effective reads obtained were highest for control (12,48,097) and lowest for heat (7,73,763). The number of OTUs obtained were 1089, 1007, 808 and 511 for control, crude, iodopropane and heat, respectively. The rarefaction curves and OTU numbers indicated that the control showed richest diversity followed by crude (slightly less than control), iodopropane and heat (Fig 4.13). Richness parameters like chao1, Shannon and Simpsons indices have indicated that the richness levels of control and crude is almost twice the richness of heat (Table 4.3).

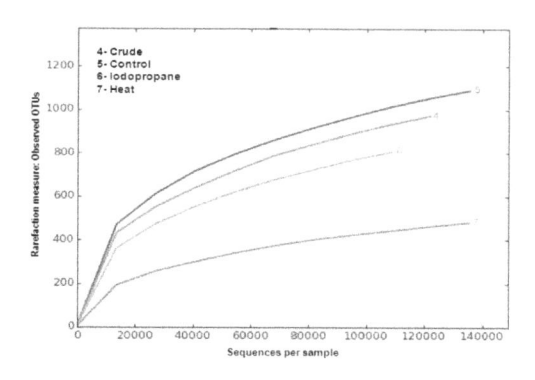

Chapter 4   Development of Electrogenic Consortia by Selective Enrichment

Fig 4.13: Rarefaction curves constructed using Unifrac.

Table 4.3: Comparison of various diversity parameters of 16s rRNA sequences obtained by MiSeq.

Sample	ACE	Chao1-lower bound	Chao1-Upper bound	Shannon	Simpson	Chao1	Observed Species
Crude	1510.837	1385.056	1670.74	5.2056	0.9145	1507.93	1007
Control	1537.387	1392.487	1616.7916	5.8661	0.948	1489.22	1089
IDP	1255.176	1128.178	1403.8295	4.9132	0.9156	1244.77	808
Heat	782.884	696.3221	914.1116	3.524	0.831	784.323	511

Principal component analysis (PCA) was done using Unifrac service (Fig 4.14). PCA analysis are in support of the other richness parameters, high diverse samples control and crude have clustered as group where as selectively enriched and low diverse pretreated samples viz., iodopropane and heat have clustered together. Unclassified sequences were observed in the all the samples which increased from the phylum to genus level. Unclassified sequences which cannot be assigned to any known taxa were high at the genus levels with comparatively high number in pretreated samples (heat- 64.28%; iodopropane- 61.06%) compared to control (58.31%) and crude (48.20%) inoculum. This is an indicative of enrichment of novel bacteria in pretreated reactors compared to control.From the Figure 4.15, it can also be seen that the phylum proteobacteria has dominated all the fuel cell with highest in iodopropane (82.17%) followed by heat (81.50%) and control (46%) among the sequences classified at phyla level while crude showed domination of firmicutes (43.79%) followed by proteobacteria (32.51 %).

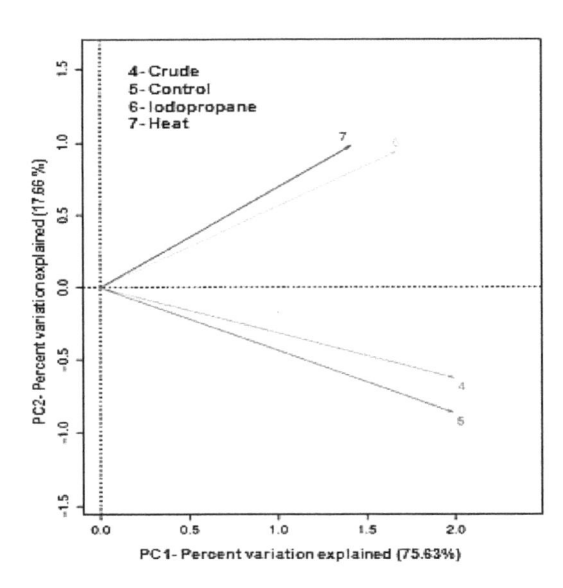

Fig 4.14: Principal component analysis (PCA) for the 4 samples studied.

Though both iodopropane pretreated and heat pretreated samples have similar community structure with respect to phyla, community structure difference with respect to various sample is only seen at the lower classification levels. This is especially true in case of proteobacteria (Fig 4.16).

4.4.2.2 Diversity of communities

The number of OTUs (808) in Iodopropane pretreated inoculum was less than crude and control indicating the selection or enrichment of only few species and inhibiting the growth of competing bacterial groups like methanogens. It is mainly composed of Beta (37.94%) and *Gamma proteobacteria* (43.93%) (phylum *proteobacteria*). Among the G*ammaproteobacteria, unclassified* genera belonging to order *Xanthomonadaceae* was most abundant (24.17%) followed by *Pseudomonadales* (4.95%). *gamma proteobacteria* like *Xanthomonas, Pseudomonas* (0.6%) and *Aeromonas* (0.51%) were known to be exoelectrogenic bacteria and were found in MFCs under various operating conditions (Yu et al., 2015). Betaproteobacterial community mainly consisted genera belonging to *Achromobacter* (11.94%) and *unclassified Alcaligens* (22.93%) of family *Alcaligenaceae* (order *Burkholderiales*). *Prevotella* (4.2%) was next abundant genus belonging to the phyla *Bacteroidetes*. Several Bacterial species belonging to *Alcaligens* and *Prevotella* were known

Chapter 4   Development of Electrogenic Consortia by Selective Enrichment

to be exoelectrogens identified in MESs operated using glucose (Li et al., 2011, Sun et al., 2010).

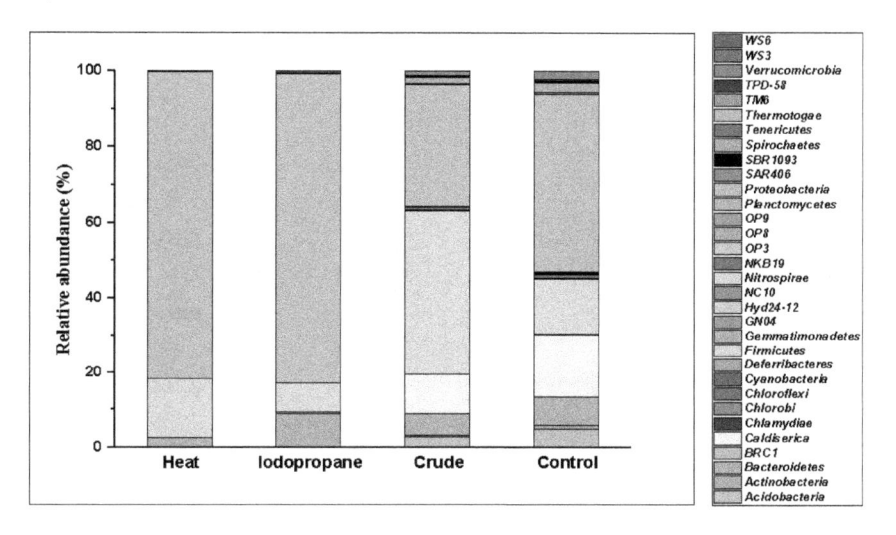

Fig 4.15: Relative abundance of various phyla in heat, Iodopropane, crude and control samples.

Firmicutes were the third dominant phyla showing an abundance of 7.9%. Dominant genera of this group include *Bacilli* (0.35%) and *Lactobacillus* (0.28%). Heat pretreatment has resulted in lowest number of OTUs (511), because of the fact that, only few bacteria can survive the 80°C.

Like Iodopropane, heat has also showed the abundace of *proteobacteria* (81.5%) followed by *firmicutes* (15.83%) and *bacteroidetes* (2.41%). Heat pretreatment showed the highest percentage of *firmicutes* among all the MESs operated. This is because of the endosporulating ability of many *firmicutes*. In heat, 90% sequences in *proteobacterial* phylum were represented by *gammaproteobacteria* and with *betaproteobacteria* filling the rest. Whereas in iodopropane, the distribution of these groups in equal. Unclassified *Xanthomonas* (45%) and *Pseudomonas* (16%) along with *Stenotrophomonas* (11%) were dominant in *gammaproteobacteria*. *betaproteobacteria* was dominated by achromobacter which constituted almost 5.809% of total reads. *Deltaproteobacteria* which represents the groups like *Delsulfovibrio, Geobacter* was only represented by small percentage of sequences. Among the *firmicutes, Lactobacillus* is represented by 12.16% sequences of total sequences while other bacilli members were represented by 1.37%. *Megasphera* belonging to

Clostridales constituted about (0.47%) while *Bacteriodes* were represented by the genera *Prevotella* (1.58%). Bacteria belonging to *Megasphera* have been shown to enrich in MESs operating with glucose (Goud and Venkata Mohan., 2013).

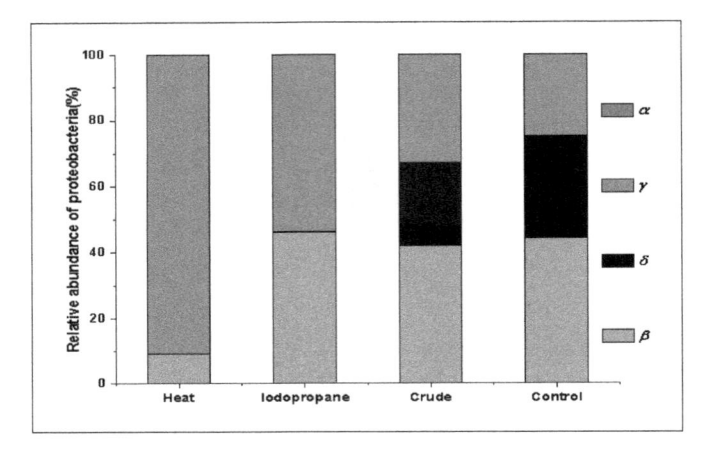

Fig 4.16: Abundance of various classes of *proteobacteria* in all the four samples and their distribution.

Control and crude inoculum has high species richness indicating that that they have a diverse group of bacteria. When diversity of these inoculums was studied at phylums level, it can be understood that there are many groups which do not have any exoelectrogenic role. This might explain the low coulombic efficiency of control. This is true particularly with the case of bacteria belonging to phylum like *Spirochaetes, Caldiserica* etc. Despite no reports available on the exoelectrogenic activity for these bacteria they were still enriched in control (*Spirochaetes*: 1.65 to 2.77; *Caldiserica*: 10.56 to 16.97) while they are completely eliminated in both the pretreated samples i.e., iodopropane and heat. Some of the known exoelectrogenic bacteria, *Shewanella sp.*, (Pirbadian et al., 2014) and *Geobacter* sp., (Reguera., et al., 2006) were found to be 0.3% and 0.15%, respectively of total abundance of iodopropane but was negligible in heat (0.001%; 0.095%), while control and crude has no sequences relating to *Shewanella* sp., and negligible for *Geobacter* sp. When diversity of pretreated inoculum is compared with crude sludge and control, the richness or number of genus present in control and crude were high. Control has shown highest number of phyla among the 4 MESs (32) where as crude has shown second highest number of phyla (28) with iodopropane showing 22 and heat showing the least i.e., 15. The reduction in number of phyla can also be considered as direct evidence for selection of only few species from crude inoculum by pretreatment.

Heat pretreatment has only selected least numbers of species because of its non-specificity unlike iodopropane which specifically inhibits particular metabolism. It should also be noted that the most of sequences at genera level could not be assigned any classification which might be due to the enrichment of novel species in the pretreatment. High bioelectrogenesis in case of Iodopropane can also be due to the fact that it is an inhibitor for methanogens (Fig 4.17). Methanogens are considered as main competitors for the several anaerobic processes like acidogenesis, fermenatation etc., and exoeletrogenesis is not an exception for them. So, eliminating them from the reactor can give huge advantage to the overall system performance which is why Iodopropane reactor has showed higher performance compared to heat and control MFC.

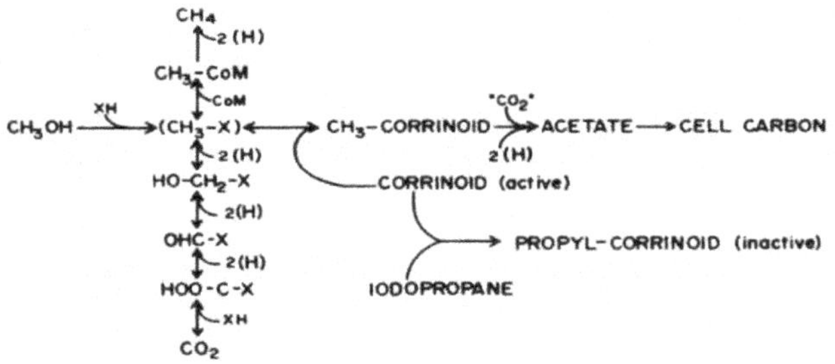

Fig 4.17: Pathway for methane synthesis in methanogens and its inhibition by Iodopropane (Kenealy et al., 1981).

4.4.2.2 Conclusions

In this study, microbial diversity of heat and iodoproapne pretreatment in samples was done to understand the structure of bacterial community enriched. Shift of microbial metabolism in pretreated biocatalyst towards electrogenesis can be attributed to shifts in microbial community. Untreated MFC depicted relatively diverse community comprising of electrogenic and non-electrogenic bacteria. While the pretreated MFC depicted enrichment of exoelectrogens which is the reason for enhanced power output in case of Iodopropane and heat MFCs. Current study illustrates the efficiency of pretreatment to enrich the exoelectrogenic bacteria for fuel cell applications which helps to deliver higher and stable power outputs.

Chapter 4   Development of Electrogenic Consortia by Selective Enrichment

4.4.3 Microbial dynamics - BESA and acid pretreatment in comparision heat and Iodopropane enrichment

In order to understand the effect of pretreatment on structure of microbial community, high throughput Miseq platform was used. V3-V4 region of 16s rRNA was used for the sequencing in this study. In order to gain insights into the changes occoured in microbiota and to compare them with the previously studied pretreatment methods, diversity analysis was performed by including microbial diversity data of Iodopropane and heat pretreatment data from the previous study.

4.4.3.1 Species richness indicators

In order to understand the effect of pretreatment on structure of microbial community, high throughput Miseq platform was used. V3-V4 region of 16s rRNA was used for the sequencing in this study. In order to gain insights into the changes occoured in microbiota and to compare them with the previously studied pretreatment methods, diversity analysis was performed by including microbial diversity data of Iodopropane and heat pretreatment data from the previous study. The number of OTUs obtained after preprocessing of sequences were 1089, 1007, 808, 511, 727 and 476 for control, parent, Iodopropane, heat, acid and BESA, respectively. The rarefaction curves and OTU numbers indicated that control and parent inoculum showed highest diversity followed by pretreatment systems Iodopropane, acid, BESA and heat (Fig 4.18). Richness parameters like Shannon and Simpson's indices, chao1 have indicated that species richness of control and parent inoculum is higherthan pretreatment systems (Table 4.4).

Table 4.4: Species richness parameters of various pretreatment methods compared to control BES and parent inoculum.

Samples	ACE	Shannon	Simpson	Chao1	Chao1 lower bound	Chao1 upper bound	Observed species
Acid	1183.299	4.321	0.890	1225.333	1090.232	1410.684	727
Parent	1510.837	5.206	0.915	1507.930	1385.056	1670.740	1007
Control	1537.388	5.866	0.949	1489.223	1392.488	1616.792	1089
Iodo-propane	1255.176	4.913	0.916	1244.775	1128.179	1403.830	808
Heat	782.885	3.524	0.831	784.323	696.322	914.112	511
BESA	756.154	3.576	0.781	752.231	663.428	883.107	476

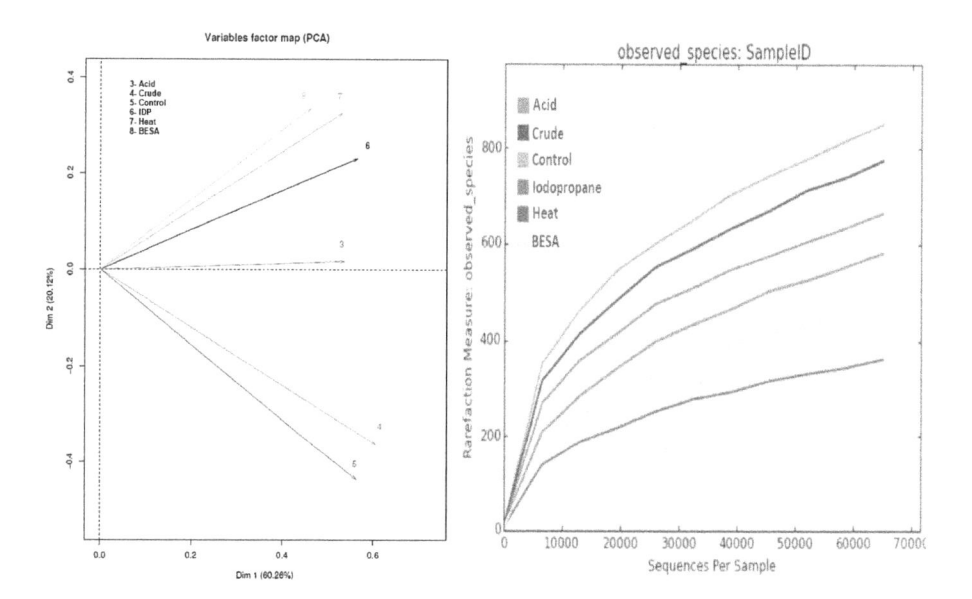

Chapter 4 Development of Electrogenic Consortia by Selective Enrichment

Fig 4.18: Principal component analysis (PCA) and rarefraction curves for the 4 pretreated MES reactors compared to control and parent inoculums.

In case of heat and BESA, chao1 indices are half the value of either control or parent indicating the drastic reduction in diversity in pretreatment reactors. Principal component analysis (PCA) was done using Unifrac service (Fig 4.18) and is in support of the other species richness parameters. Samples with high diversity (control and parent samples) have clustered as single group while the low diverse pretreated samples viz., BESA, iodopropane and heat have clustered together. Acid pretreated sample has not clustered with any of the groups indicating its unique diversity. Unclassified sequences were observed in the all the samples which cannot be assigned to any known taxa were high at the genus levels with comparatively high number in pretreated samples indicating enrichment of novel bacteria in pretreated reactors compared to control.

From the Figure 4.19, it can also be seen that the phylum *proteobacteria* has dominated in all the MESs with highest in iodopropane (82.17%) followed by heat (81.50%), acid (81%), BESA (48.80 %) and control (46%) among the sequences classified at phyla level. Parent (43.79%) and BESA (42.674%) samples have shown highest incidence of *firmicutes*.

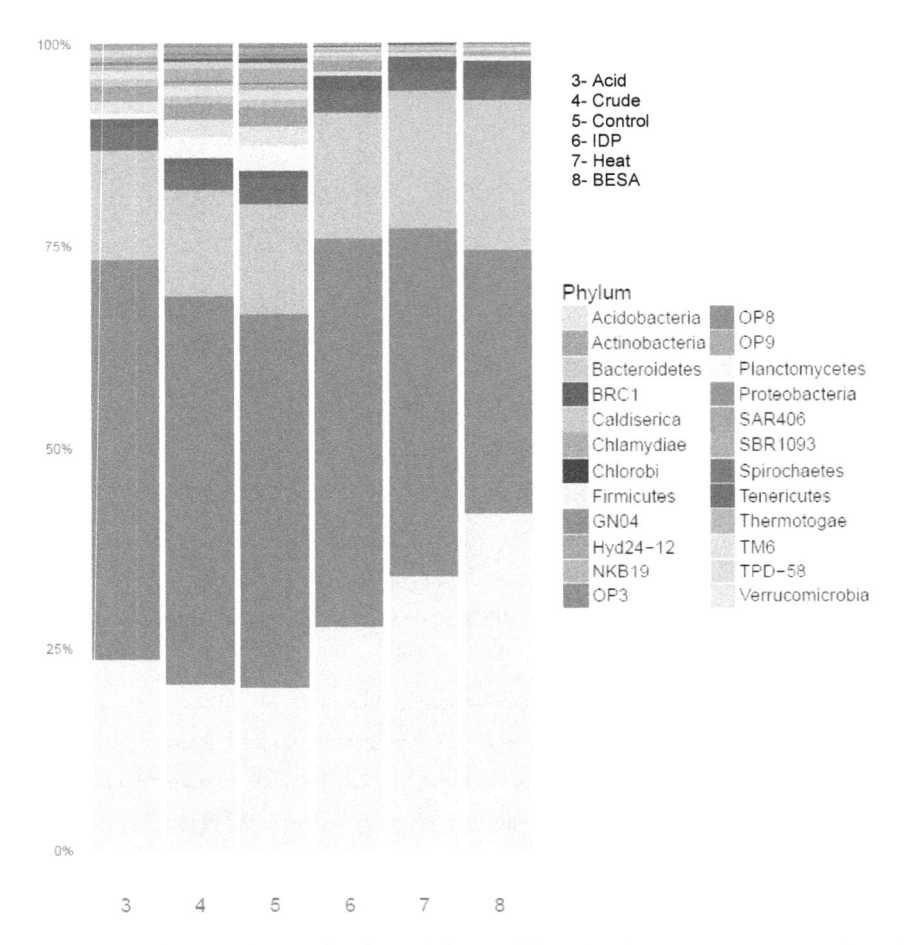

Fig 4.19: Relative abundance of various phyla in acid, BESA, heat, Iodopropane, crude and control samples.

4.4.3.2 Diversity of communities

The number of OTUs in acid pretreated inoculum was observed to be less when compared toparent inoculum and control indicating the selection or enrichment of only few species and inhibiting the growth of competetivenon-exoelectrogenic bacterial groups. Among the phylum *proteobacteria,* gamma (63.002%) and beta (17.63 %) *proteobacteria* were found to be dominant. Among the *gammaproteobacteria, unclassified* genera belonging to order *Xanthomonadaceae* was most abundant (43.34%) followed by *Pseudomonadales* (9.45%). *Gamma proteobacteria* like *Xanthomonas, Pseudomonas* and *Aeromonas*were already known

Chapter 4   Development of Electrogenic Consortia by Selective Enrichment

to be exoelectrogenic bacteria and were found in several MFCs (Yu et al., 2015). *Beta proteobacterial* community mainly consisted genera belonging to *Achromobacter* (13.85%) followed by *Alcaligens* (Fig 4.20).

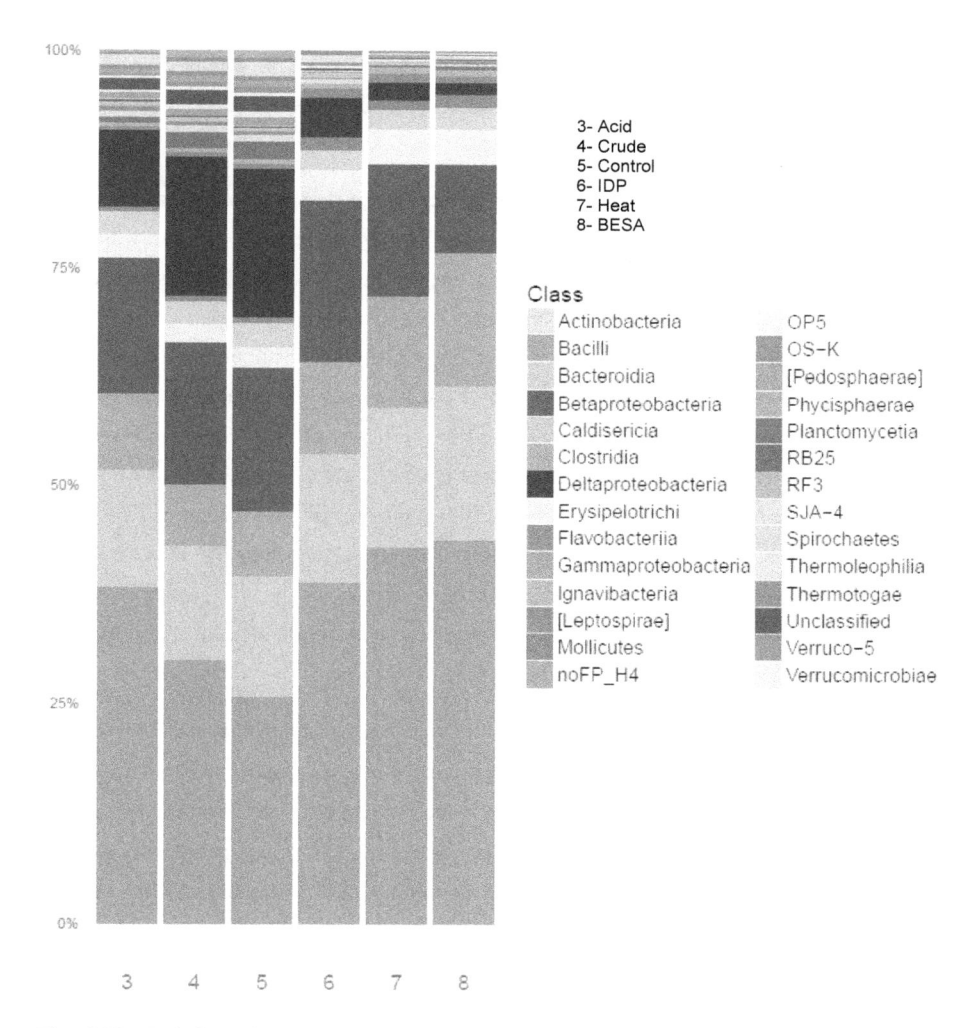

Fig 4.20: Relative abundances of various classes of in all the six samples and their distribution.

Several Bacterial species belonging to *Alcaligens* and *Prevotella* were known to be exoelectrogens identified in microbial fuel cells operated using glucose (Li et al., 2011, Sun et al., 2010). Similar trend was observed in case of Iodopropane BES which has showed good bioelectrogenic performance in previous study. However, the main difference between acid and Iodopropane enrichment is the increased incidence of bacteri belonging to the order *Pseudomonaceae* in acid compared to Iodopropane reactor which explain why acid MES was able to produce highest bioelectrogenic activity compared to iodopropane as *Pseudomonas* are known be electrogenic and its mechanism has been studied in detail. Like acid and iodopropane, heat has also showed the abundace of *proteobacteria* (81.5%) followed by *firmicutes* (15.83%) and *bacteroidetes* (2.41%).

Heat pretreatment showed the second highest percentage of *firmicutes* among all the MFCs operated. This is because of the endosporulating ability of many *Firmicutes* which can survive the harsh conditions applied during pretreatment. In heat, 90% sequences in proteobacterial phylum were represented by *Gammaproteobacteria* and with *betaproteobacteria* filling the rest. Whereas in iodopropane, the distribution of these groups in equal. Unclassified *Xanthomonas* (45%) and *Pseudomonas* (16%) along with *Stenotrophomonas* (11%) were dominant in *gammaproteobacteria*. *Betaproteobacteria* was dominated by achromobacter which constituted almost 5.809% of total reads. Among the *firmicutes, Lactobacillus* is represented by 12.16% sequences of total sequences while other bacilli members were represented by 1.37%. *Megasphera* belonging to *Clostridales* constituted about (0.47%) while *Bacteriodes* were represented by the genera *Prevotella* (1.58%). Bacteria belonging to *Megasphera* have been shown to enrich in MFCs operating with glucose (Goud and Venkata Mohan., 2013).

Control and crude inoculum have high species richness indicating that that they have a diverse group of bacteria. When diversity of these inoculums was studied at phylums level, it can be understood that there are many groups which do not have any exoelectrogenic role. This might explain the low coulombic efficiency of control. This is true particularly with the case of bacteria belonging to phylum like *Spirochaetes, Caldiserica* etc. Despite no reports available on the exoelectrogenic activity for these bacteria they were still enriched in control (*Spirochaetes*: 1.65 to 2.77; *Caldiserica*: 10.56 to 16.97) while they are nearly eliminated in both the pretreated samples i.e., iodopropane and heat.

Some of the known exoelectrogenic bacteria, *Shewanella sp.,* (Pirbadian et al., 2014) and *Geobacter* sp., (Reguera., et al., 2006) were found to be 0.3% and 0.15%, respectively of total abundance of iodopropane but was negligible in heat (0.001%; 0.095%), while control and crude has no sequences relating to *Shewanella* sp., and negligible for *Geobacter* sp. When diversity of pretreated inoculum is compared with crude sludge and control, the richness or number of genus present in control and crude were high. Control has shown highest number of phyla among the all the MESs operated (32) where as crude has shown second highest number of phyla (28) with iodopropane showing 22, acid showing 21 and heat and BESA showing the least i.e., 15 and 14, respectively. The reduction in number of phyla in pretreated MES can also be considered as direct evidence for selection of only few species from crude inoculum by pretreatment. While in case of untreated control diversity has been preserved even after several cycles of operation. Heat and BESA pretreatment has only selected few species owing to their growth inhibition. It should also be noted that the most of sequences at genera level could not be assigned any classification which might be due to the enrichment of novel species in the pretreatment.

Changes in functioning in a new environment will depend in part on ecological changes in the composition and relative abundance of species. These changes can be natural or artificially induced like in case of pretreatment. Some species will grow better in new conditions than others, and poorly growing species might go extinct (Steudel et al., 2012; Mori et al., 2013). Functioning will then depend on the traits of the surviving species after ecological sorting rather than the original species complement. Changes in functioning might also depend on evolution of traits of constituent species in response to selection from new environmental conditions or from shifts in the surrounding biotic community. Evolution might improve ecosystem functioning. For example, evolution of enhanced growth of constituent species, on average, would increase community productivity. In addition, selection for greater specialisation and niche partitioning could increase the range of resource use and total growth of the community (Gravel et al., 2011; Lawrence et al., 2012).

4.4.3.3 Conclusions

The experimental results of microbial diversity analysis using MiSeq platform has demonstrated that the pretreatments have clear impact on the community structure of the mixed inoculums. Such impact of selective enrichment has resulted in enhanced growth of constituent species i.e, electrogens which increase the productivity of the total community wiith respect to the function of MES. In addition, selective enrichment has resulted in greater specialisation and niche partitioning which can increase the range of resource use or enhanced electrogenic activity. It can be understood from diversity studies that *proteobacteria* are the most dominant group of bacteria among all the enriched incoulcums followed by *firmicutes*. In case of both acid and BESA, it can be observed that electroactive bacterial groups like*Pseudomonadales* and *xanthomonodales* are the more dominant. Among the *firmicutes* enriched in BESA and heat pretreatment MFCs, *Bacilli* and *Clostridia* are the most dominant which were also known Exoelectrogens. While the untreated control and parent inoculms have shown relatively high diversity when compared to pretreatment samples. It can be observed that the in untreated control there are several groups enriched which has no role electrogenesis which can explain the low performance of control MFC. Overall, this study has given insight into community structures of pretreated inoculms and helped in understanding the functionality of respective pretreatment method.

4.5 Microbial electrochemical systems for bioelectricity production with simultaneous treatment of paper and pulp wastewater

Paper and pulp industry has become one of the largest industries in the world due to the huge demand for paper and its related products. During the paper manufacturing process, wastewater with intense color, characterized by the presence of chlorides used for bleaching, organic content in the form of cellulose, lignin, etc. (Pokhrel and Viraraghavan, 2004) is generated. The composition and organic content of paper and pulp wastewater is varying and depends on the pulping process. Conventional wastewater treatment methods were currently employed for treating paper mill effluent which includes physical, chemical and biological methods (Buyukkamaci and Koken, 2010). Many recalcitrant pollutants still exist after treatment by conventional biological and chemical treatments together. Therefore, focus on new treatment methods like electrochemical treatment is in progress (Anglada et al., 2009). Although electrochemical treatment is efficient in the removal of highly recalcitrant pollutants, its energy intensive nature prevents its use. Alternate methods are being explored where a renewable source could supply this energy like in casr of microbial electrochemical systesms. The main advantage of using a MES system is that the external supply of voltage is not necessary, unlike in electrochemical treatment techniques where input of energy is obligatory. The *in situ* biopotential developed in the system due to the presence of electrode assembly and the substrate degradation by biocatalyst is sufficient to replenish the energy input. In this study an attempt was made to enumerate the potential of single chambered MES to treat real field paper and pulp wastewater with simultaneous power generation. Integrating coagulation along with MES for achieving further treatment efficiency was also studied.

4.5.1 Experimental

4.5.1.1 Paper Mill Effluent

Real-field effluent acquired from paper mill unit was used in this study. The wastewater has a characteristic dark brownish color [Color, 4000 Hazen units) with chemical oxygen demand (COD) concentration of 4100 mg/l (pH, 6.1; total dissolved solids (TDS), 2500 mg/l; total suspended solids (TSS), 500 mg/l; phosphates, 155 mg/l; sulfates, 241 mg/l; volatile fatty acids(VFA), 950 mg/l]. Effluent prior to use was stored at 4°C till further usage.

4.5.1.2 Biocatalyst

Sludge from anaerobic digester was used as parent inoculum in both the biosystems. Before starting the reactors, the inoculum was enriched under anaerobic microenvironment at pH 6.0 for several cycles in designed synthetic wastewater (DSW) with glucose as carbon source [glucose- 3 g/l; NH_4Cl - 500 mg/l, $MnCl_2$ - 15 mg/l, KH_2PO_4-250 mg/l, K_2HPO_4- 250 mg/l, $CuCl_2$-10.5 mg/l, $MgCl_2$ - 300 mg/l, $CoCl_2$ -25 mg/l, $ZnCl_2$ -11.5 mg/l, $CaCl_2$- 5 mg/l, $FeCl_2$ - 25 mg/l, $NiCl_2$ -16 mg/l].

4.5.1.3 MFC system design and operation

To know the efficiency of MES (as microbial fuel cell (MFC)) in producing power and treating paper and pulp wastewater in comparison to anaerobic treatment (AnT), two dimensionally similar bioreactors were designed with perspex material with a total/working volume of 700/500 ml. The first system i.e. single chambered microbial electrochemical (MES) system was fixed with non-catalyzed graphite plates (5 cm^2, 1 cm diameter) as electrodes. Anode was kept submerged in the anolyte (wastewater) where as cathode was exposed to air (air cathode). Proton exchange membrane (Nafion, Sigma Aldrich) was sandwiched between the anode and cathode to facilitate only the exchange of the proton. No mediators were used for this study. Second system i.e. anaerobic reactor (AnT) was operated as control without electrode-membrane assembly to assess the relative efficiency of the MES in treating wastewater. Both systems were operated in fed batch mode at room temperature under anaerobic conditions. Magnetic stirrer (100 rpm) was used to mix the anolyte throughout the operation and to prevent the settling of the inoculum. Both the systems were operated with DSW initially, which was later substituted with real field paper mill effluent once the system depicted stable performance. Once stabilized performance with respect to bio-electrogenic activity was noticed, both the systems were shifted to paper mill effluent and operated at COD load of 3000 mg/l. Initially both the reactors were operated with paper and pulp effluent at initial pH of 6 for acclimatization and stabilization of acidogenic inoculum and prevent the growth of non-electrogens and subsequently shifted pH of 7. pH of the feed was adjusted using concentrated orthophosphoric acid and 0.1 N sodium hydroxide. Only inlet pH of wastewater was set to desired pH and change in pH was monitored with time. The treated effluents from AnT and MES systems were further studied for color removal employing alum (0.1%) and $FeCl_3$ (0.015 %) as coagulation aids. After adding coagulants to treated wastewater, pH was adjusted to 6. After subjecting the contents for stirring continuously for 10 minutes at 50 rpm, the contents were allowed to settle and the supernatant obtained was used for analysis. Both systems were operated at organic load adjusted to 3000 mg/l using distilled water.

4.5.1.4 Analysis

Power output parameters viz., voltage and current (with 100 Ω resistor) were continuously monitored using digital multimeter. Polarization behavior of MES system was studied with the function of current density against potential and power density under varying external resistances (30 KΩ to 50 Ω). Anode potentials along with cell *emf* were measured across various resistances (30 KΩ to 50 Ω). Wastewater parameters like chemical oxygen demand (COD), volatile fatty acids (VFA), nitrates, phosphates, sulphates, color and pH were analyzed from time to time according to the standard methods (APHA, 1998). Bio-electrochemical characteristics of the MES was studied with the potentiostat-galvanostat system (PGSTAT12, Ecochemie) and all the obtained data was analyzed using GPES version 4.9 software (AUTOLAB). Experiments were carried out by considering anode and cathode as working and counter electrodes, respectively with reference to the Ag/AgCl(S) electrode. Cyclic Voltammograms were obtained by ramping potential on working electrode (anode) between -0.5 V to 0.5 V at a scan rate of 30 mV/s. Chronoamperometry was performed by applying a potential of 500 mV on working electrode for 300 seconds and sustainable current generation by the system was analyzed.

4.5.2 Results and Discussion
4.5.2.1 Bioelectrogenic activity- MFC Vs AnT
4.5.2.1.1 Chronoamperometry (CA)

Chrono-amperometry was performed at the stable phase of operation to enumerate the sustainable current generation in both MES and AnT systems by applying a potential of 500 mV for 300 seconds (Fig 4.21). During CA operation, maximum current generation will be observed which will show a gradual drop and stabilize at some point which can be considered as sustainable current over a period of time (Chandrasekhar and Venkata Mohan, 2012). Higher sustainable currents were observed with MES system (10 mA) in comparison to AnT (2 nA), attributing to the influence of existing electrode assembly in the MES system that induce a potential difference between the biocatalyst and electrodes. It is reported that MES system favors the growth of exo-electrogens by providing electrogenic environment for bacteria which triggers the release of more reducing equivalents onto the electrode resulting more catalytic currents and higher redox currents.

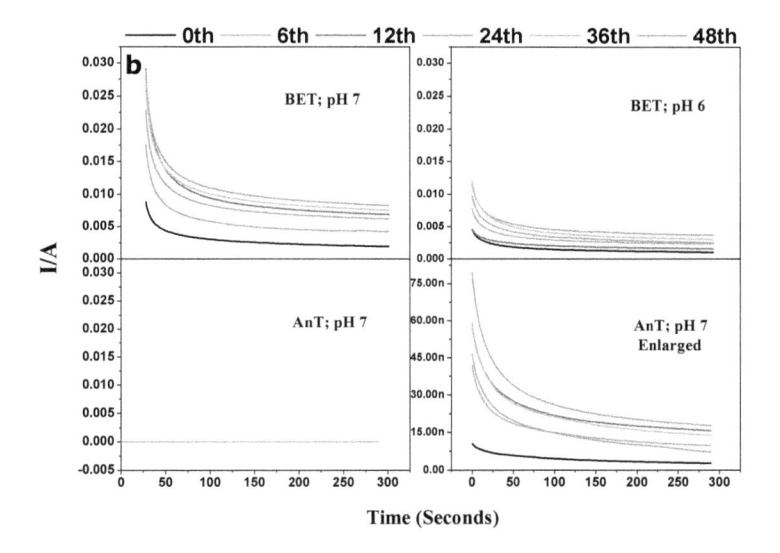

Fig 4.21: Chrono-amperometry depicting the production of sustainable current during the MES and AnT operation.

In addition, MES operation at pH 6 and pH 7 showed a significant variation in the stable current generation. Operation at pH 7 resulted in higher sustainable currents of 11 mA, whereas pH 6 operations resulted in 4 mA of stable current generation. Operation of MES at different pH yielded difference in bio-electrogenic activity because of the variation in pH that altered both growth and biochemical characteristics of the biocatalyst. Finally, MES operation at pH 7 was found to be optimum in the present study with paper and pulp as substrate.

4.5.2.1.2 Cyclic voltammetry (CV)

Voltammograms (cyclic) documented clear distinction between MES and AnT systems in the form of oxidation currents (OC) and reduction currents (RC) indicating the difference in the behavior of biocatalyst (Fig. 4.22). The redox catalytic currents were observed to increase at each time interval irrespective of the systems (MES and AnT). However, higher redox catalytic currents of (OC: 8 to 28 mA; RC: -2 to 1 mA) were recorded with MES operation, whereas AnT operation documented lower catalytic currents of (OC: 11.7 to 80 nA; RC: -80 to 40 nA).

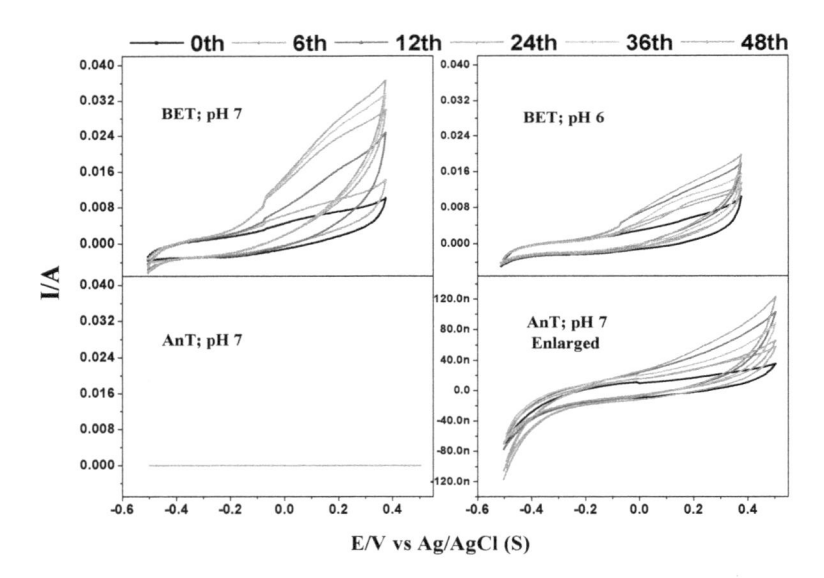

Fig 4.22: Cyclic voltammograms showing variation in the electron discharge pattern (+0.5 to -0.5 V) with anode and cathode as working and counter electrodes against Ag/AgCl (S) reference electrode (scan rate, 30 mV/s).

Specifically with MES, oxidation sweeps were higher than reduction sweeps (RC was almost same in both the phases of operation) while AnT operation showed higher reduction currents compared to oxidation currents. Higher oxidation currents in MES represent, higher oxidation of substrate in anode chamber and higher availability of reducing equivalents i.e., electrons and protons for the production of bioelectricity. Reduction currents were higher in AnT system during the entire cycle operation, which might be due to the absence of electrodes in AnT system that might have favored for reduction pathways. In the case of MES, highest catalytic currents were generated at 24th hour and minimum by 48th hour indicating the exhaustion of substrate. Minor variations in catalytic currents with time were observed in the case of anaerobic operation (at a scale of nA).

Besides the comparative difference between MES and AnT, the operation of MES at different pH resulted in significant difference in terms of redox catalytic currents. MES at pH 7 documented highest catalytic currents (OC: 36.7 mA; RC: -6.45 mA) in terms of both oxidation and reduction sweeps indicating higher metabolic activity of the biocatalyst at pH 7 whereas operation at pH 6 showed comparatively lower performance (OC/RC, 19.7/-4.65 mA) than pH 7 (Fig 4.22). COD removal was observed to be higher at pH 7 in comparison to pH 6 resulting in liberation of higher number of redox equivalents that will yield high bioelectrogenic activity which is in support with the observed redox catalytic currents from CV.

4.5.2.1.3 Electron Transfer

Tafel analysis is used to understand and analyze the bioelectro-kinetic parameters in terms of redox slopes (b_a and b_c) and behavioral shifts of the biocatalyst during the experimental operation (Chandrasekhar and Venkata Mohan., 2012; Raghavulu et al., 2012). Higher oxidation slopes indicate the requirement of higher activation energy and results in lower oxidation. Similarly, lower oxidation slope requires lower activation energy and results in higher oxidation. The same can be related in the case of reduction slopes as well. In addition, polarization resistance (R_p) can also be calculated from the Tafel slopes which provide information on the resistance to electron transfer. Higher redox slopes were observed with AnT (ba: 0.186V/dec; bc: 0.137 V/dec) operation in comparison to MES (ba: 0.103 V/dec; bc: 0.007 V/dec) which indicates the requirement of high activation energy for electron transfer in AnT system than MES due to the lack of electrode assembly and electroactive consortia (Fig 4.23). In addition, the comparative difference between MES operation at pH 6 and 7 showed a significant variation in redox slopes. pH 7 operation resulted in low oxidation slope during initial hours (0.083 V/dec) followed by an increment at 24th h (0.15 V/dec) which again showed a drop by the end of operation (0.103 V/dec). This indicates the rapid substrate degradation capabilities during initial hours followed by electron losses which might be due to the occurrence of methanogenic environment that would have utilized the free electrons towards other metabolic pathways.

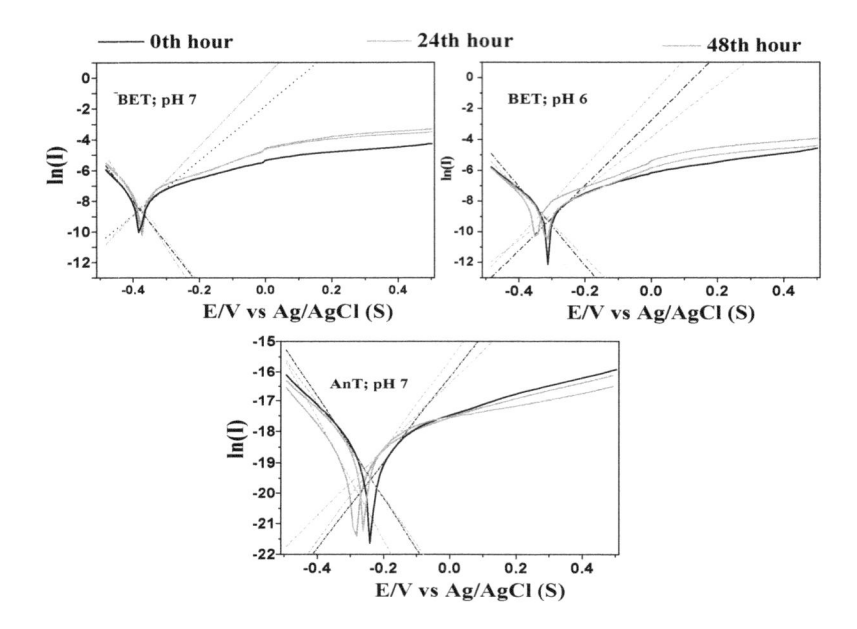

Fig 4.23: Tafel slopes derived with the function of biosystems, pH and time.

The decrement in slope at the end of operation indicates the efficiency of metabolic activity of biocatalyst. The reduction slope was high during the initial hours (0.129 V/dec) followed by a continuous decrement till the end of operation (0.07 V/dec) indicating electron neutralization/reduction reactions towards power generation and pollutant degradation. Operation at pH 6 favored oxidation reactions depicting low oxidation slope till 24th h (0.1 V/dec) which again increased by the end of operation (0.137 V/dec). This implies that the electrons might have been utilized towards volatile fatty acid production. The reduction slope was similar to oxidation slope which showed an increment by the end of operation (0.104 V/dec) attributing to the lowered neutralization efficiency of biocatalyst at pH 6.

Polarization resistance (R_p) is observed to be higher in AnT (2.97 kΩ) than MES (19.29Ω) signifying the impact of electrode assembly in MES that allows the regulated electron transfer through the electrodes. R_p was nearly 10^6 times higher in AnT than MES system which indicates the lack of electrogenic environment and electrochemically active bacteria due to the absence of electrode assembly. In addition, MES at pH 7 showed low R_P (13.29 Ω) in comparison to pH 6 (28.1 Ω). The influence of pH on the growth and metabolic activity of biocatalyst would have resulted in the difference in electrogenic activity at both operations where the acidogens enriched at pH 6 would have functioned well at pH 7 taking the advantage of rapid substrate degradation capabilities during pH 7 operation.

4.5.2.1.4 First derivative

Derivatives of cyclic voltammograms obtained during various time intervals of MES operation were scanned for the involvement of redox couples as peaks and the obtained peak potentials (E_{peak}) (against Ag/AgCl, KCl (3.5M) reference electrode) were modified with respect to standard hydrogen electrode (SHE). These corrected values were correlated with standard redox potentials ($E°_{Redox}$) of various metabolic reactions or redox couples occurring in bacteria. Most of the redox couples identified were associated with metabolic pathways occurring either on bacterial membrane like Kreb's cycle or electron transport chain (ETC). In pH 6 operation, several E_{peaks} were identified corresponding to each time period (Fig. 4.24). E_{Peak} at -0.010 V was consistent during entire cycle corresponding to redox couple FAD/FADH$_2$ and also the pyruvate dehydrogenase which have NAD$^+$/NADH(H$^+$) couple as its co-enzyme (Metzler., 2003).

Other E_{Peaks} observed during operation were near to 0.220 V, 0.05 V, 0.300 V, 0.130 V and 0.350 V. E_{Peak} near to 0.220 V can be attributed to succinate dehydrogenase (SDH; 0.222 V), ubiqinone/semiquinone (CoQ/UQH- 0.211 V) where as E_{peak} near to 0.130 V corresponds to FAD/FAD(H) (0.131 V) activity in its bound form to enzymes. Other E_{peaks}~0.300 V and 0.350 V corresponds to cytochromes involved in the electron transport chain and finally E_{peak} ~0.05 V can be related to enzyme glutamate dehydrogenase (0.061 V) which has NAD/NADH as its coenzyme. Operation at pH 7 also revealed E_{peak} at -0.010 V which was constant during entire cycle. Other E_{peaks} identified at 6th and 12 hour of operation were near to 0.20 V corresponding to SDH and CoQ/UQH redox couple. SDH is an important membrane bound enzyme involved in both bacterial metabolism and ETC. It uses the bound FAD/FADH$_2$ couple to complete the reaction. CoQ/UQH is mobile carrier present in the plasma membrane and involves in bacterial electron transport. Throughout the 24th, 36th and 48th hours of operation, two E_{peaks} were observed i.e. at -0.01 V and 0.20 V. E_{peak} at 0.20 V can be related to various enzymes using bound flavin (FAD) as coenzyme which holds the electron for certain period and delivers to other electron carrier. In case of AnT system, no peaks were observed during operation.

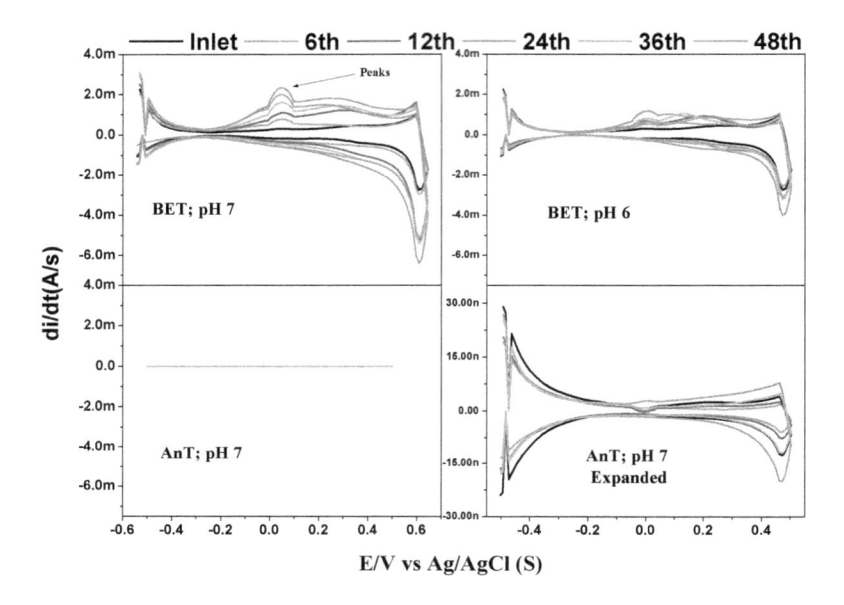

Fig 4.24: Derivative cyclic voltammograms (DCV) recorded during operation of MES and AnT systems with respect to different pH and time.

During MES operation, various membrane bound enzymes like glutamate dehydrogenase, α-ketoglutaric acid, succinate dehydrogenase, pyruvate dehydrogenase, malate dehydrogenase etc. were identified. These enzymes either have a free or a bound coenzymes like $NAD(P)^+$, FAD(H), FMN(H), etc. which help in catalyzing the redox reactions at the membrane interface making them probable sites for exo-electron transport. Introduction of an artificial electron acceptor (electrode) in MES system might have showed positive effect by up regulating the expression of these redox proteins on the bacterial membrane. In addition to these enzymes, few redox couples like CoQ/UQH and cytochromes which are involved in electron transport were also identified. Few reports claim that the coenzyme compounds like flavins (FMN, riboflavin and riboflavin-5-phosphate) can shuttle between bacteria and electrode making process more efficient (Marsili et al., 2008).

Chapter 4 Development of Electrogenic Consortia by Selective Enrichment

4.5.2.2 Fuel cell behaviour
4.5.2.2.1 Bioelectrcity

Apart from considerable wastewater treatment efficiency, MES as microbial fuel cell (MFC) documented effective power output. During the operation with DSW, MES showed increment in power output with every additional cycle and showed a maximum OCV of 323 mV at 192th hour and 7.2 mA/m^2 of current density during the 5th hour of operation (4th cycle) (Fig 4.25).

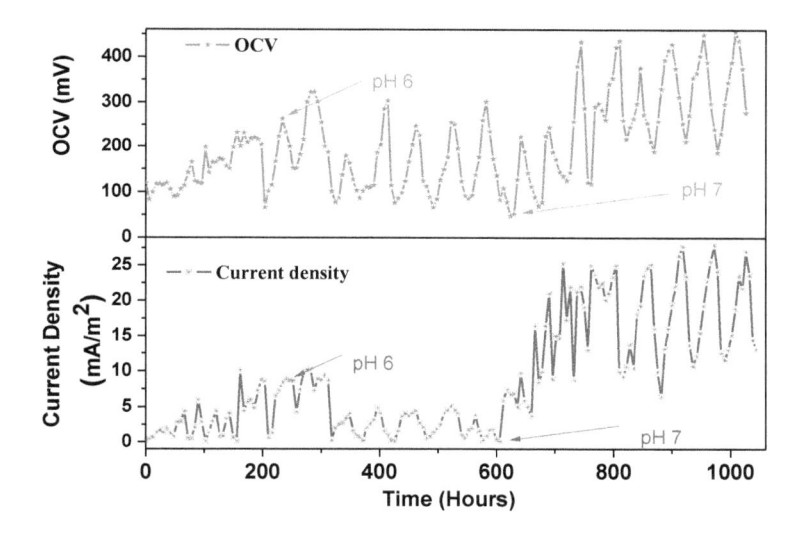

Fig 4.25: Progressive increase in OCV and current density generation during MFC operation with paper and pulp wastewater.

After attaining the stabilized performance, the reactor fed with paper and pulp wastewater and was operated at pH 6 and 7. Initially, the reactor was operated at pH 6 which usually favors the growth of acidogens and enriches them selectively. Maximum OCV and current density were observed to be 302 mV and 10.28 mA/m^2, respectively during 12th cycle. Though the current density was higher during the initial phase of operation, a decrement was observed in the later phase which showed a stabilized performance for the next 6 consecutive cycles with slight variations in current density and OCV. Bioelectrogenic activity and substrate removal efficiency of fuel cell was found to increase with time and improvement of power output was observed after replenishing the fuel cell with fresh substrate every time. As there was a drop in current density during the operation at pH 6, MES was operated at pH 7 and the performance was monitored regularly where the system showed a maximum OCV and current density of 456 mV at 1008th h and 28 mA/m^2 at 972nd h, respectively. The enriched acidogens would have functioned well at neutral pH taking the advantage of rapid substrate consumption favored at pH 7 which might be the result for observed higher electrogenic performance. Stabilized performance in terms of OCV and current density was observed at pH 7 unlike pH 6 and similarly, maximum power density was observed at pH 7 (40.2 mW/m^2) compared to pH 6 (6.03 mW/m^2).

4.5.2.2.2 Polarization

Polarization curve is a key parameter to access the performance of both the chemical and microbial based fuel cells and provides insights into the losses incurring in the system during the operation. At pH 6 operation, cell design point (CDP) or maximum power density (MPD) was observed at 100 Ω with power density of 6.03 mW.m^{-2} (Fig 4.26).

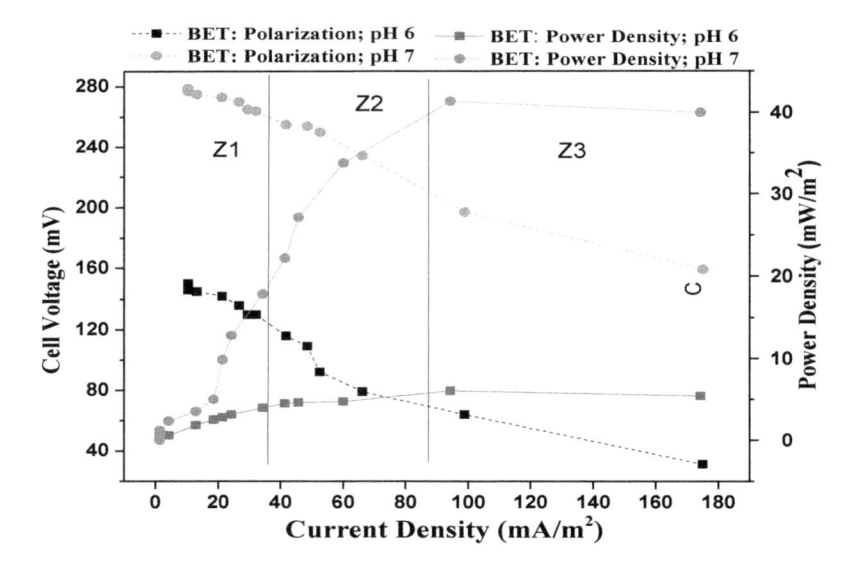

Fig 4.26: Polarization curves (Z1 is zone of activation losses; Z2 is zone of ohmic losses; Z3 is zone of concentration losses) of MFC operated at pH 6 and pH 7.

After shifting to pH 7, CDP is observed at 200 Ω, with maximum power density of 41.37 mW/m^2. Increase in both CDP and power density enumerates the efficiency of MES operation at pH 7 which favored the metabolic activity of the acclimatized acidogens during neutral pH. For an ideal system, the polarization curve would be a straight line as there will be no losses at any point. While in the case of MFC, polarization plot will not be a straight line as there are various losses that can be incured during the transfer of electrons from substrate to the final electron acceptor (Zhao et al., 2009; Venkata Mohan et al., 2014b). These losses can be divided into activation losses (Z1) which originate from factors like microbial metabolism where the slow irreversible reactions of bacterial metabolism make bacteria to spend extra energy on producing electrons and transferring them to electrodes which results in voltage losses. Activation losses are characterized by steep decrease voltage at low currents. During pH 6 operations there is very rapid decrease in voltage till 10 Ω while in the case of pH 7, voltage drop was gradual which indicates the decrease in activation losses with time. Ohmic losses (Z2) which in the form of linear drop at increasing currents occur due to resistance of electrode distance between electrodes, proton exchange membrane, resistance to ions in the electrolyte, etc.

With long term operation of MFC, decrease of ohmic losses was evident from pH 6 to 7 operations where a more linear voltage drop was observed at pH 6 than the operation at pH 7. Finally the concentration losses (Z3) or mass transfer losses, which occur mainly due to the limiting of transport of substrate to biocatalyst and further to electrodes, removal of products from the site and electron acceptor conditions at cathode. It is characterized by drop of voltage at near maximum currents. Concentration losses get reduced with time in MES operation. With long term operation of MES with paper and pulp wastewater as feed, losses showed consistent reduction which might be due to the acclimatization of microbial community to the system environment (Fig 4.26).

4.5.2.2.3 Cell potentials

Anode (E_{anode}), cathode ($E_{cathode}$) and cell potentials (E_{cell}) are important parameters which control the performance of a fuel cell. Anode potential (E_{anode}) is very important for regulating the interaction between biocatalyst and anode. It dictates the maximum amount of free energy (ΔG) available for bacteria to grow. When E_{anode} is at higher potential, bacteria can gain more free energy from the oxidation of the organic compounds (i.e. by donating electrons to the anode).At the same time for cathode to function properly, there should be a maximum potential difference between the two electrodes. So, E_{anode} should be precisely balanced so that there will be optimum potential for the flow of electrons from bacteria to cathode. Anode and cathode potentials were measured across resistance ranging from 30 kΩ to 50 Ω. Anode potential varied from MFC- pH 6 operation (E_{anode} -349 mV) to MFC- pH 7 (E_{anode}-312 mV) operations. E_{anode} of MFC- pH 7 operations showed more positive potential when compared to MES pH 6 (Fig 4.27) which indicates that the faster growth of bacteria in MES- pH 7 system due to the availability of more free energy for bacteria. Cathode potential was constant during the entire process indicating that the anode (artificial electron acceptor) was the deciding factor for the dissimilarity in performance of this system.

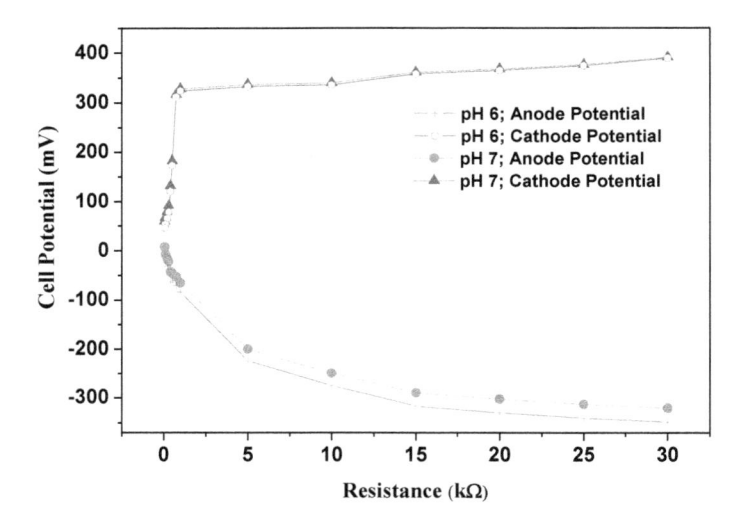

Fig 4.27: Cell potentials and relative decrease in anodic potential (RDAP) profile drawn against varying external load during stabilized performance of MFC with the function operating pH.

4.5.2.2.4 Relative decrease in anode potential (RDAP)

The short circuit current generated without or with a random resistor is high which exists for a very short period of time and quickly reaches minimum. To prevent this, a suitable resistance has to be applied between anode and cathode such that there will not be any current limiting conditions. RDAP was performed to know this effective resistance, by step wise decrease in the resistance from 30 kΩ to 50 Ω. Cell potential and current with respect to given resistance was measured (Menicucci et al., 2006). MES-pH 6 operation has illustrated higher external sustainable resistance (14 kΩ) which decreased in MES-pH 7 to 9 kΩ (Fig 4.28). This decrease in optimum resistance for sustainable current production indicates that the kinetic (of electron transfer) and mass transfer limitations of MES decreased making system efficient (Raghavulu et al., 2012).

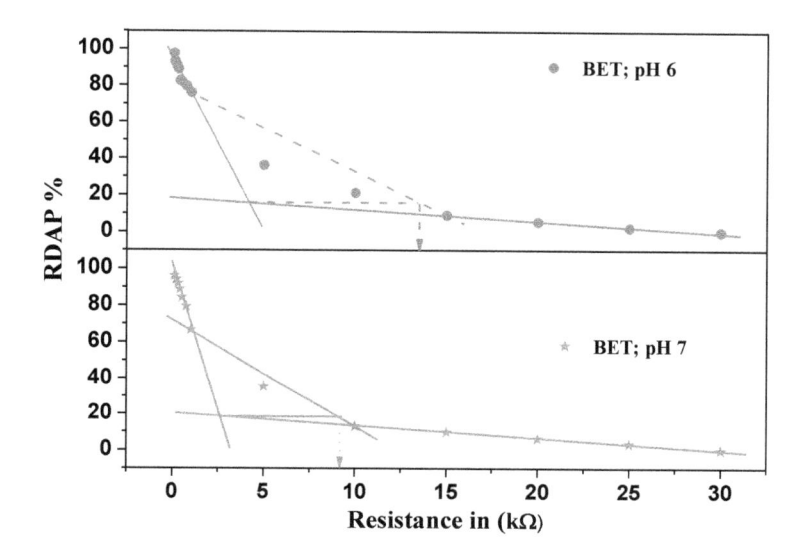

Fig 4.28: Relative decrease in anodic potential (RDAP) profile drawn against varying external load during stabilized performance of MFC with the function operating pHs.

4.5.2.3 Bioremediation-MESVsAnT

MES showed relatively higher substrate (COD) degradation efficiency compared to the AnT operation at pH 6 (Fig 4.30). During the initial cycles of operation both MES and AnT showed nearly similar COD removal efficiencies (MES, 32%; AnT, 29%). With an increase in time COD removal efficiency was observed to increase in MES system (MES, 45%; AnT 34%), which indicates the acclimatization of electrochemically active bacteria (EAB) with increase in number of cycle operations. The presence of electrode assembly in MES showed enhanced anodic electron transfer reactions, which might have also played major role in enhancing the substrate degradation (Venkata Mohan et al., 2009; Velvizhi and Venkata Mohan, 2011). EAB transfer the electrons at a faster rate leading to effective degradation of organic substrate than the anaerobic bacteria. COD removal efficiency was observed to be higher in MES (59%) at pH 7 than MES-pH 6 (43%) and AnT (49%) operation (Fig 4.30).

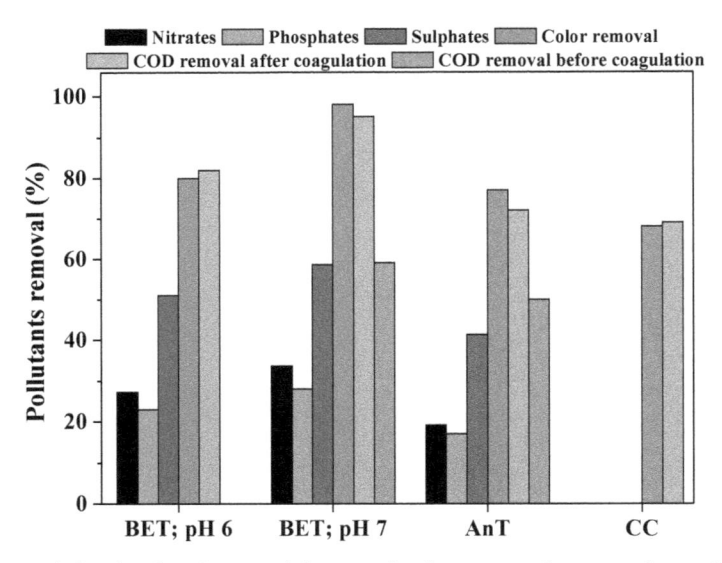

Chapter 4 — Development of Electrogenic Consortia by Selective Enrichment

Fig 4.29: Variation in phosphates, sulphates and colour removal pattern observed with the function of time during the operation of MES and AnT systems (HRT, 48 hrs; organic loading rate, 3000 mg/l).

In addition to substrate degradation, the performance of biosystems was also evaluated for their capacity to remove multipollutants viz., nitrates, phosphates and sulphates. Many pollutants present in the wastewater act either as a source of minerals for microbial growth or as electron donors in anodic reactions making MES a very good option for waste remediation (Kelly and He, 2014; Venkata Mohan et al., 2014a). MES operation illustrated higher removal efficiency of nitrates (33%) than AnT (19%) operation (Fig 4.29). Since nitrate is also an effective terminal electron acceptor in the absence of oxygen, the nitrates get reduced to nitrogen. Moreover reduction of nitrate also generates proton motive force for effective mobility of protons. Nitrates get reduced by both assimilative and dissimilative processes which require five electrons to get converted to nitrogen (Virdis et al., 2008). In the case of phosphates, maximum removal was observed at pH 7 (33.54 %) followed by pH 6 (27.28 %) when compared to the anaerobic process (17%). With increase in pH up to 7.8 enhanced uptake of phosphates by bacteria was reported (Wang et al., 2013; Liu et al., 2007). Nitrates and phosphates also serve as nutrients in the metabolic activities. Sulfates, removal was observed to be higher in MES operation at pH 7 (58.59 %) in comparison to anaerobic control (41.03%). In MES operation, the self-induced bio-electrocatalytical conditions facilitate sulphate reduction. Based on the thermodynamic hierarchy under limited oxygen availability conditions, sulphates can also act as terminal electron acceptors (Dutta et al., 2008). In a MES system, oxidation and reduction of sulphates occurs concurrently in the anodic chamber with simultaneous reduction of organic matter present in the wastewater. The elemental sulphur gets deposited on the anode and acts as a mediator for transferring the electrons (Dutta et al., 2008). Specifically, the results obtained illustrate the efficacy of pH 7- bioelectrogenic microenvironment in wastewater remediation.

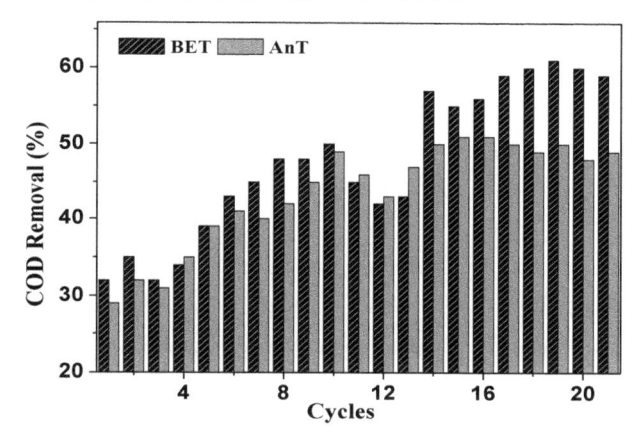

Fig 4.30: Variation in COD observed with the function of time during the operation of MES and AnT systems (HRT, 48 hrs; organic loading rate, 3000 mg/l).

4.5.2.4 Integrated process for color removal

Lignin and its derivatives are primarily responsible for the brown color in paper and pulp wastewater. The effluent from the conventional biological treatment process contains a high content of color, lignin and COD (Fig 4.31). In the present study, chemical coagulation (CC) was employed to further remove color and COD from the treated effluents of MES and AnT operation. A control setup with crude effluent was used for comparison.

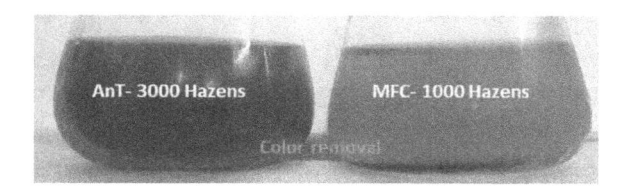

Fig 4.31: Color removal efficiency of MFC and AnT reactors.

Chemical coagulation has been used to remove pollutants i.e. colloids and ions from wastewater through destabilizing the suspensions facilitating them to precipitate and permitting their removal through sedimentation or filtration (Venkata Mohan et al., 1998; Ahmad et al., 2008). Coagulation experiments were performed by using combination of alum ($KAl(SO_4)_2$; 1 g/l) and ferric chloride ($FeCl_3$; 0.15 g/l) by adjusting the pH of aqueous phase to 6 and stirring for 10 min (25 rpm). Further the settled supernatant was analyzed for the removal of color and COD. Coagulated outlet from MES-pH 6 operation showed good color (80%) as well as COD (82%) removal efficiencies. The outlet from MES system with pH 7 operation documented complete removal of color along with high COD removal efficiency (95%). AnT effluent showed lower removal efficiency (color/COD, 77/72%). Comparatively chemical coagulation alone showed lowest performance (color/COD, 68/69%).

The pollutants present in the wastewater exist in the form of organic/inorganic (colloids) and heavy metals (ions) which have electrical charge. These colloidal compounds can be destabilized by the reduction of their net surface charge (repulsive potential of the electrical double layer and accumulation of counter ions) through the addition of oppositely charged ions (as coagulants). In chemical coagulation, the van der waals forces hold the colloids together and allow them to aggregate and settle. Experimental data specifically illustrated the color and COD removal efficiencies of MES in comparison to the corresponding AnT and chemical coagulation alone. MES is a hybrid bio-electrocatalyzed process combining electrode setup (electrode membrane assembly) with an anaerobic system. The induction of an artificial electron acceptor i.e. anode makes it a potential site for many bio-electrochemical reactions (direct anodic oxidation reactions-DAO) and secondary physico-chemical reactions (indirect anodic reactions- IAO) induced by the *in situ* generated potential, which substantially increase the treatment efficiency of the MES system (Venkata Mohan et al., 2008; Velvizi et al., 2014 ; Chandrasekhar and Venkata Mohan, 2012). The microenvironment in MES plays a crucial role for bio-electrolytic oxidation, which also facilitates generation of more charged ionic species that might also have shown a positive influence on the integrated coagulation process than the AnT integrated processes. Due to the formation of these charged ionic species, the destabilized colloidal particles formed after MES treatment might have further aggregated to form larger flocs. Due to this the destabilization of colloidal particles after MES treatment might aggregate color colloidal compounds to larger flocs. This integrated process closely resembles electro-coagulation generally used for wastewater treatment but without applying external potential.

4.5.2.5 VFA and pH variation

pH is critical in governing the metabolic pathway and depending on the organism and growth condition, changes in external pH can bring about alterations in several primary physiological parameters, including internal pH, concentration of other ions, membrane potential and proton-motive force (Slonczewski et al., 2009). With operation, pH dropped in both the systems (Fig 4.32). At 4th hour the pH in both the reactors was almost same (5.8) but from 8th hour of operation pH drop was higher in AnT (5.5) compared to MES (5.7).

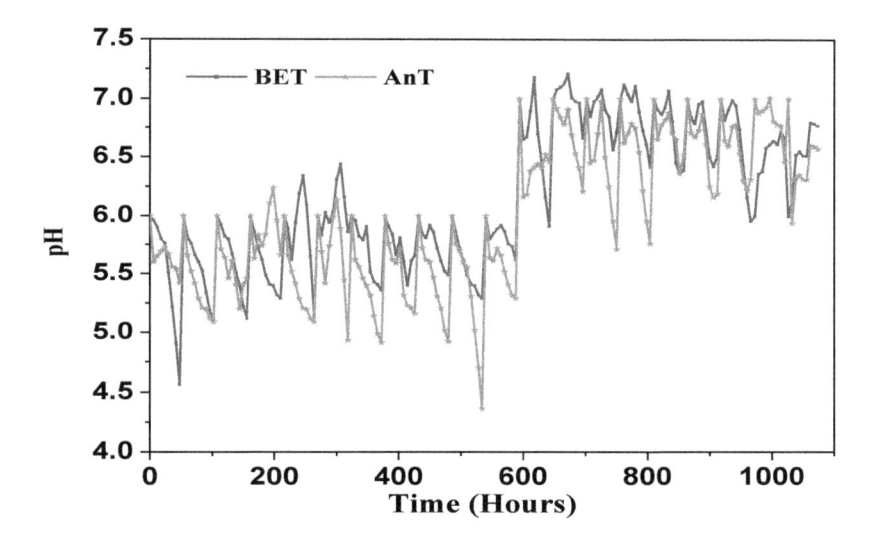

Fig 4.32: Variations of pH documented during MFC and AnT system operation at with the function of pH and time

In general, VFA production in a system influences the pH. VFA increased from 550 mg/l to 720 mg/l in MES and 755 mg/l in AnT at 4th hour. 1800 mg/l of VFA was observed at 24th hour in AnT whereas only 1356 mg/l was observed in MES which lead to a drop in pH to 5.1 and 5.3, respectively. The drop in outlet pH and VFA accumulation was more in the case of AnT (pH, 4.5; VFA, 2600 mg/l) compared to MES (pH, 4.8; VFA, 2100 mg/l). In the second phase both the reactors were operated at pH 7. Even electricity generation was high at pH 7 operation. From 12th hour VFA concentration increased from 890 mg/l to 2300 mg/l in AnT dropping pH to 5.7. After 30th hour VFA consumption resulted in increased pH. VFA enumerates extent of the acidogenic activity in anaerobic degradation process (Fig 4.33).

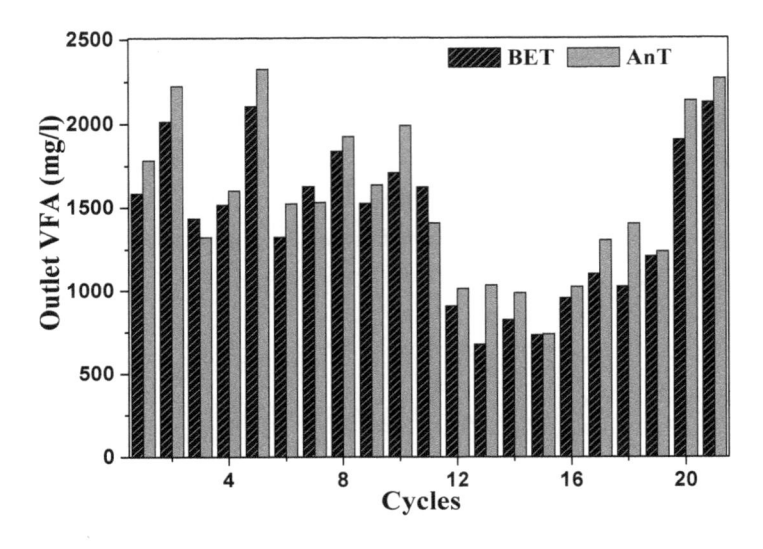

Fig 4.33: Variations of VFA documented during MFC and AnT system operation at with the function of pH and time.

VFA and pH are integral expressions of the acid-base conditions of any anaerobic process as well as intrinsic index of the balance between two of the most important microbial group's viz., acidogenic and methanogenic (Goud and Venkata Mohan., 2012). Operation of MES under acidophilic conditions (pH 5.5 and 6) helps to limit the methanogenic bacterial activity. Acidophilic pH around 6 compared to a near neutral pH can increase conversion efficiency of the substrate to H^+ and e^-. Neutral pH limits the H^+ concentration in the solution but acidophilic pH is favorable for the H^+ transfer in Between anode and cathode. Moreover, acidophilic pH favors growth of acidogenic bacteria (Venkata Mohan et al., 2007) that leads to acidogenic pathway during degradation of substrate generating H^+ and e^- to make power. pH range of 5.5–6 was considered to be ideal to avoid both methanogenesis and solventogenesis (Raghavulu et al., 2009; Venkata Mohan et al., 2007).

4.5.3 Conclusions

Microbial electrochemicalsystem (MES) is one of the emerging technologies with good waste remediation/mineralization capacity and a potential to generate energy simultaneously. MES showed good COD, color and multiple pollutants removal when compared to AnT. CV and DCV analysis showed the involvement of membrane bound proteins, which can increase the bioelectrogenic activity. MES illustrated higher catalytic current than AnT. The bioelectrogenic activity was further supported by RDAP, polarization and cell potential analysis. The work illustrates that integration of MES system with chemical coagulation can be another credible approach for effective treatment of complex wastewaters eliminating the use of external load.

CHAPTER 5
Summary and Future Perspectives

5.1 Summary and Conclusions of the Research

Microorganisms which perform extracellular electron transport are the key to various applications of microbial electrochemical systems (MES). The electrochemical interactions between the microorganism and electrodes is a complex event and gaining insights in these interactions can lead to improved performance of MES and also help in real field implementation. Even though, many electroactive microorganisms have been discovered till date, knowledge of EET is well known only in few bacteria like *Geobacter, Shewanella* and *Pseudomonas*. With more and more bacteria being assigned to be electroactive, gaining insights into their mechanisms and their role is important for understanding electroactive communities, their ability to form biofilms and electroactivity.

To enhance the extracellular electron transport and increased carbon flux, *ndh* gene coding NADH dehydrogenase II (NDH-2) from *Bacillus subtilis* was cloned into *E. coli* and successful transformation of *E. coli* was confirmed using colony PCR. NDH-2 protein was expressed using 1M IPTG at 37°C which was purified using Ni-NTA affinity chromatography and confirmed by western blotting (Anti-his antibodies). To evaluate the metabolically engineered strain's extracellular electron transfer ability, iron reduction assays were performed with ferric citrate as final electron acceptor. Engineered strain expressing NDH-2 has shown higher iron reduction capability compared to the parent strain. Increase in total NADH pool and $NAD^+/NADH$ ratio in NDH-2 cells indicates that expression of NDH-2 has showed increased metabolic activity (carbon utilization) which can be related to the higher reduction of ferric citrate. To understand the role of NDH-2 on electron flux in MES, NDH-2 strain was used as biocatalyst for the MES operation. From the Chronoamperometry studies at +200 mV it can be seen that the bioelectrogenesis has increased up to 3-fold in engineered strain when compared to the parent strain. EIS has indicated that the MES operated with NDH-2 has showed decrease in charge transfer resistance which could be because of increased biofilm formation and also increased electron flux.

Further study of NDH-2 for biohydrogen production has showed that its expression can result in increase in hydrogen production. This study has showed that studying the proteins involved in native electron transport can also provide detailed understating of their role on extracellular electron transport and can be studied to increase the bioelectrogenic activity. This study has successfully demonstrated that expression of NDH-2 can increase the overall performance of microbial electrochemical systems by enhancing metabolic and electron flux. Further electrochemical studies with purified protein and techniques like protein film voltammetry and spectro-electrochemistry can give insights into electrochemical behavior of NDH-2 its reaction kinetics which can be applied for optimizing the potentials that can used for enhancing biofilm formation on anode or cathode or both and to increase electron flux in microbial electrochemical systems.

In the second objective it was revealed that the electroactivity or columbic efficiency of the mixed culture operated MES can be increased by selectively enriching electroactive bacteria or by eliminating competing non-electroactive bacteria. In order to do this, various inoculum pretreatment strategies were evaluated. In the first sub-objective, we have used the heat and Iodopropane (IDP) as methods of inoculum pretreatment. From the results it was observed that the Iodopropane inoculum pretreatment has resulted in the highest electrogenic activity when compared to heat and control. Bioelectrogenic activity can also be correlated to the other bioelectrochemical parameters like cyclic voltammetry and Chronoamperometry where IDP showed highest redox currents and sustainable currents, respectively followed by heat pretreated MES and control. Columbic efficiency, which gives information on amount of substrate converted to electricity, has indicated that the IDP MES has highest percentage of electrogenic activity which can also be correlated to the percentage of electrogenic bacteria present in the total mixed inoculum.

In order to know the microbial diversity of enriched inoculums, 16s rRNA sequencing was performed using MiSeq platform which has showed that Iodopropane and heat pretreated inoculums showed enriched and less diverse microbial community. Richness parameters have indicated that the diversity in enriched inoculums has decreased compared to the parent inoculum and control MES. Decrease in species richness indicates that inoculum pretreatment has exerted selective pressure on certain groups of bacteria which are either favourable for the exoelectrogens or negative pressure on the undesirable non-electrogenic bacteria. Closer look at the diversity has indicated that the Iodopropane pretreatment has resulted in high relative percentage of phylum proteobacteria which has highest percentage of exoelectrogens discovered till now. Heat pretreatment has resulted in enrichment of bacteria belonging to phyla Firmicutes which are endospore formers when subjected to stress. Among the phyla Firmicutes, bacteria belonging to Clostridia were found to be enriched which are also known to be exoelectrogens.

To further enhance the bioelectrogenesis of mixed biocatalyst for MES application, pretreatment strategy was extended to other two methods viz., acid pretreatment (acid shock) and BESA pretreatment which were already proven to be excellent pretreatment strategies for biohydrogen production. BESA is a potent methanogenic inhibitor and is specifically used in this study to inhibit the growth of methanogens which are unwanted group of bacteria for bioelectrogenesis. Among the two pretreatment methods used, acid shock turned out to be the best strategy to enrich electrogenic bacteria as it has shown highest bioelectricty generation and substrate to bioelectricity generation efficiency. Acid pretreated reactors have shown VFA production, which resulted in lower pH drop compared to the BESA and control MES indicating the lower fermentation activity in acid pretreated MES compared to control where fermentative metabolism is more prevalent over electrogenic metabolism. Excessive formation of VFA in control reactor also explains the decrease in columbic efficiency where most of the substrate utilized by biocatalyst was converted to metabolic acids.

In order to understand the composition of microbial community, high throughout sequencing using Miseq platform was performed. To understand the effectiveness of the acid shock and BESA pretreatment over previous studies, diversity of acid shock and BESA samples were compared to previously pretreatment studies (Iodopropane and heat). Microbial diversity analysis has showed the domination of electrogenic bacteria in pretreated MFCs, while untreated control and parent MFC has shown more diverse community indicating that pretreatment has resulted in enrichment of electroactive bacteria. Except in the case of heat pretreatment where *firmicutes* were found to be dominant, in all the other enrichment methods, bacteria belonging to phylum *proteobacteria* were found to be dominant. Overall, this work has illustrated the efficiency of pretreatment to enrich the exoelectrogenic bacteria in MFC which is also a cost-effective method.

Microbial electrochemical system was used for treatment of real field paper and pulp wastewater in comparison to the anaerobic treatment. It was observed that MES reactor showed high organic contaminants and multiple pollutants removal when compared to anaerobic treatment. Color removal after respective treatment and chemical coagulation was observed to be higher in MES compared to anaerobic treatment. Bioelectrogenic parameters like cyclic voltammetry and derivational cyclic voltammetry analysis showed the involvement of several membrane bound proteins in the bioelectrochemical reactions, indicating direct electron transfer. MES also showed higher catalytic current than AnT indicating higher electrogenesis. The bioelectrogenic activity of MES was further supported by RDAP, polarization and cell potential analysis. The work illustrated that integration of MES system with chemical coagulation which can be another futuristic approach for effective treatment of complex wastewaters eliminating the use of external load. In addition to this, MES can be successfully operated with no or very low power input which makes it sustainable and environmentally friendly system.

5.2 Future perspective

From the overall work, it can be understood that the performance of MES can be significantly enhanced by metabolic engineering of a single strain or by biofilm engineering of mixed culture using pretreatment strategies. It is very certain that, to enhance performance of microbial electrochemical systems, be it for bioelectricity generation, bioelectrosynthesis, or bioelectrochemical treatment, systems biology approach combined with synthetic biology is a perfect tool. Synthetic biology can be used to improve already existing metabolic networks or circuits and combining them with novel engineered genetic circuits. Synthetic biology tools can be used to increase the electron flux or redirect the excess electron flow into the extracellular electron transport there by enhancing the efficiency of MES. As our knowledge of the intricate genetic and metabolic networks with respect to extracellular electron transport increases, more avenues will be opened up for rational engineering of microorganisms with superior performance in MES. By using metabolic engineering, one can address factors affecting the current generation in MFC, from enhanced degradation of organic fraction to improved transfer of electrons and communication with electrodes.

Bioelectrogenic performance of exoelectrogenic bacteria in mixed cultures significantly depends on the other non-electrogenic bacteria. Establishing synergistic interactions among them can generate higher power densities and higher substrate degradation. Mixed microbial populations, can adapt to process disturbances, environmental distress and with no need for sterilization, and can consume wide range of organic substances. Therefore, development of mixed microbiome-based processes is important to increase the efficiency of electricity and to expand future real field applications of MES. There is still lot of information that needs to be studied about microbial community assembly and the microbial dynamics of the interactions between complex organic substrates degrading communities and the electroactive communities. From the inoculum pretreatment studies, it can be understood that bacterial interactions in mixed culture can be improved by pretreating the mixed culture before inoculating into MES. This can potentially improve the syntrophic associations among the bacteria. Using artificially or naturally assembled communities is another approach for obtaining genetic and metabolic pathway diversity in a MES for long term operations for real field applications.